HEALTHCARE ACTIVE LEARNING

HAL

BIOCHEMISTRY

Start date

Target completion date

Tutor for this topic

Contact number

Using this Workbook

The workbook is divided into 'Sessions', covering specific subjects.

In the introduction to each learning pack there is a learner profile to help you assess your current knowledge of the subjects covered in each session.

Each session has clear learning objectives. They indicate what you will be able to achieve or learn by completing that session.

Each session has a summary to remind you of the key points of the subjects covered.

Each session contains text, diagrams and learning activities that relate to the stated objectives.

It is important to complete each activity, making your own notes and writing in answers in the space provided. **Remember this is your own workbook—you are allowed to write on it.**

Now try an example activity.

ACTIVITY

This activity shows you what happens when cells work without oxygen. This really is a physical activity, so please only try it if you are fully fit.

First, raise one arm straight up in the air above your head, and let the other hand rest by your side. Clench both fists tightly, and then open out your fingers wide. Repeat this at the rate of once or twice a second. Try to keep clenching both fists at the same rate. Keep going for about five minutes, and record what you observe.

0289 2486

Stop and rest for a minute. Then try again, with the opposite arm raised this time. Again, record your observations.

Suggested timings are given for each activity. These are only a guide. You may like to note how long it took you to complete this activity, as it may help in planning the time needed for working through the sessions.

Time taken on activity

Time management is important. While we recognise that people learn at different speeds, this pack is designed to take 20 study hours (your tutor will also advise you). You should allocate time during each week for study.

Take some time now to identify likely periods that you can set aside for study during the week.

Mon	Tues	Wed	Thurs	Fri	Sat	Sun
am						
pm						
eve						

At the end of the learning pack, there is a learning review to help you assess whether you have achieved the learning objectives.

HEALTHCARE ACTIVE LEARNING

HAL

BIOCHEMISTRY

Patricia A. Lyne BSc PhD RGN

Reader in Nursing Research, Department of Advanced Nursing,
The North East Wales Institute

THE OPEN LEARNING FOUNDATION

CHURCHILL LIVINGSTONE
EDINBURGH LONDON MADRID MELBOURNE NEW YORK AND TOKYO 1995

CHURCHILL LIVINGSTONE
Medical Division of Longman Group UK Limited

Distributed in the United States of America by Churchill Livingstone Inc., 650 Avenue of the Americas, New York, N.Y. 10011, and by associated companies, branches and representatives throughout the world.

First published 1995

ISBN 0 443 05272 7

British Library of Cataloguing in Publication Data
A catalogue record for this book is available from the British Library.

Library of Congress Cataloging in Publication Data
A catalogue record for this book is available from the Library of Congress

Printed and bound by Bell and Bain Ltd., Glasgow

For The Open Learning Foundation

Director of Programmes: Leslie Mapp
Series Editor: Robert Adams
Programmes Manager: Kathleen Farren
Production Manager: Steve Moulds

For Churchill Livingstone

Director (NAH): Peter Shepherd
Project Development Editor: Mairi McCubbin
Project Manager: Valerie Burgess
Design Direction: Judith Wright
Pre-press Project Manager: Neil Dickson
Pre-press Desktop Operator: Kate Walshaw
Sales Promotion Executive: Hilary Brown

The publisher's policy is to use **paper manufactured from sustainable forests**

Contents

OPEN LEARNING FOUNDATION TEAM MEMBERS

Writer: Dr Patricia A. Lyne
Reader in Nursing Research, Department of Advanced Nursing,
The North East Wales Institute

Editor: Dr Elisabeth Clarke
Distance Learning Programme Coordinator,
Royal College of Nursing

Reviewer: Professor Justus Akinsaya
Pro Vice Chancellor, International Development,
Anglia Polytechnic University

Series Editor: Robert Adams
OLF Programme Head,
Social Work and Health and Nursing,
University of Humberside

The Open Learning Foundation

Higher education has grown considerably in recent years. As well as catering for more students, universities are facing the challenge of providing for an increasingly diverse student population. Students have a wider range of backgrounds and previous educational qualifications. There are greater numbers of mature students. There is a greater need for part-time courses and continuing education and professional development programmes.

The Open Learning Foundation helps over 20 member institutions meet this growing and diverse demand – through the production of high-quality teaching and learning materials, within a strategy of creating a framework for more flexible learning. It offers member institutions the capability to increase their range of teaching options and to cover subjects in greater breadth and depth.

It does not enrol its own students. Rather, The Open Learning Foundation, by developing and promoting the greater use of open and distance learning, enables universities and others in higher education to make study more accessible and cost-effective for individual students and for business through offering more choice and more flexible courses.

Formed in 1990, the Foundation's policy objectives are to:

- improve the quality of higher education and training
- increase the quantity of higher education and training
- raise the efficiency of higher education and training delivery.

In working to meet these objectives, The Open Learning Foundation develops new teaching and learning materials, encourages and facilitates more and better staff development, and promotes greater responsiveness to change within higher education institutions. The Foundation works in partnership with its members and other higher education bodies to develop new approaches to teaching and learning.

In developing new teaching and learning materials, the Foundation has:

- a track record of offering customers a swift and flexible response
- a national network of members able to provide local support and guidance
- the ability to draw on significant national expertise in producing and delivering open learning
- complete freedom to seek out the best writers, materials and resources to secure development.

INTRODUCTION

This unit is one of three closely-related units which are designed to introduce people with little or no previous experience to the biological sciences which underpin the health care professions.

Each of the three can be studied on its own. No knowledge of the others is assumed. However, if you intend to study more than one of these units, it is best to do *Genetics* first, then *Biochemistry*, and finally *Physiology*.

In this unit, I would like to introduce you to biochemistry, which is the study of the chemical processes which take place in living cells.

People are sometimes deterred from studying this subject by the thought of the chemistry and physics on which it is based. I hope to show that it is quite possible to achieve an understanding of some of the fundamental concepts in biochemistry without building on a basis of pure science. The approach will be practice related, with an emphasis on selected areas of the subject which can help to improve understanding of the provision of care to patients and clients. No prior knowledge of the subject is assumed.

Biochemistry is often called 'the chemistry of life'. It is concerned with the chemical processes which go on inside living cells. This term is not very helpful to people who don't have a background in chemistry, as it doesn't really help them to visualise what biochemistry is about.

Perhaps we can start from a different standpoint, thinking about what the individual cell is doing and how this contributes to the health of the whole body. Then we can gain an understanding of the processes which need to take place if the cell is to make its proper contribution.

The events which take place inside cells cannot readily be described in non-technical language. They are outside our normal experience. So I have tried to find a metaphor to use for what the cell is doing. The one which best suits the cell is that of a small business enterprise. This allows the various cellular processes to be described in terms that are reasonably familiar – like 'import' and 'export'.

Session One: The Cell as a Small Business

This first session explains why the metaphor of a small business seems to suit the cell so well. It introduces different kinds of cells, describes what they do and how they are fit for that purpose; finally it outlines how that purpose is carried through.

Session Two: Cellular Machine Tools

The second session examines the first stage of the manufacturing process—making things to make other things with. In the cell, this involves making special proteins. The session describes how genetic material in the nucleus controls the manufacture of proteins, and indicates some of the things that can go wrong. The section ends with a discussion about the role of health carers in maintaining these processes.

Session Three: The Cellular Machinery

Session Three considers the cell's machinery; it describes what proteins are and how one type, the enzymes, work. It looks at the conditions that enzymes need to work properly, at what can go wrong and at how this produces some of the problems we see in health care.

Session Four: Importing Raw Materials

We now look at some of the raw materials the cell needs – what they are, where they come from and how they are brought to the cell. Finally this session describes how these raw materials get through the factory gates, that is, into the cell itself.

Session Five: The Power Supply

This session outlines different types of work carried out by all cells, and some of the specialist work done by a few. It examines the power supply which makes this work possible, the efficiency of the system, and how the cell deals with by-products arising from its power supply.

Session Six: Packaging and Transport

Session Six considers what happens to the cell's products; it examines how they are moved around inside the cell and how they are exported. The session concludes by returning to one of the specialised cells used as an example throughout the unit, the β cell of the pancreas, and looking at how it packages and despatches insulin.

Session Seven: Quality Control

The final session outlines the ways in which the activities of the cell are co-ordinated and how the quality of its products are maintained. This involves consideration of some of the things that can go wrong and how this affects the health of the whole person.

You may find that, even though this unit uses non-technical language, you need to spend quite a long time on some sessions. Many people find this to be the case because biochemical concepts are rather challenging and unfamiliar. Please don't worry if you find that progress is slow. It is worth while persevering with these concepts because they are so fundamental to the understanding of how our bodies function in health and illness.

LEARNING PROFILE

Below is a list of learning outcomes for each session in this unit. You can use it to identify your current learning, and so to consider how the unit can help you to develop your knowledge and understanding. The list is not intended to cover all the details discussed in every session, and so the learning profile should only be used for general guidance.

For each of the learning outcomes listed below, tick the box that corresponds most closely to the point you feel you are at now. This will provide you with an assessment of your current understanding and confidence in the areas that you will study in this unit.

	Not at all	Partly	Quite well	Very well
Session One				
I can:				
• explain in general terms how different types of cell are adapted for their particular function.	☐	☐	☐	☐
Session Two				
I can:				
• identify the structures on which protein synthesis takes place in the cell	☐	☐	☐	☐
• describe the mechanism of protein synthesis	☐	☐	☐	☐
• explain how protein synthesis is controlled by the nucleus	☐	☐	☐	☐
• identify ways in which health care can support this function.	☐	☐	☐	☐
Session Three				
I can:				
• describe how some proteins function as enzymes	☐	☐	☐	☐
• identify what enzymes actually do	☐	☐	☐	☐
• describe the conditions that enzymes need	☐	☐	☐	☐
• explain how they can be affected by environmental and genetic factors.	☐	☐	☐	☐
Session Four				
I can:				
• describe how food materials are broken down into a relatively small number of basic materials	☐	☐	☐	☐
• explain how these materials – especially glucose – get into the cell	☐	☐	☐	☐
• describe how this range of materials can be used to make all the cell's products.	☐	☐	☐	☐
Session Five				
I can:				
• describe where the energy to drive cellular activity comes from	☐	☐	☐	☐
• explain what mitochondria are and how they function	☐	☐	☐	☐
• explain where mitochondria are thought to have originated.	☐	☐	☐	☐

	Not at all	Partly	Quite well	Very well
Session Six				
I can:				
• describe how materials get in and out of the different parts of the cell	☐	☐	☐	☐
• explain how proteins are packaged for transport when they are complete	☐	☐	☐	☐
• outline how manufactured products get out of the cell.	☐	☐	☐	☐
Session Seven				
I can:				
• explain the need for control and integration within the cell	☐	☐	☐	☐
• describe methods of internal and external control of cell metabolism	☐	☐	☐	☐
• outline the role of hormones in external and internal control	☐	☐	☐	☐
• describe the action of insulin	☐	☐	☐	☐
• explain the concept of integration as applied to metabolic pathways in the cell.	☐	☐	☐	☐

PRELIMINARY ACTIVITY

ALLOW 15 MINUTES

In *Resource 1*, you can read a personal account of the experience of diabetes. This is an illness which is caused by a disturbance of the biochemistry of human cells and we will be considering it in some detail in this unit.

Read through it now and make a list of:

- the questions which it raises in your mind
- the aspects of this person's experience which you would like to understand better.

At the end of this unit we will look at *Resource 1* again and you will be able to see how your understanding of patients with this illness and of the role of the specialist nurse has developed.

SESSION ONE

The cell as a small business

Introduction

This particular metaphor was chosen because it is the nearest I could get to finding a recognisable model for describing the cell. If we think of each cell as a unit which manufactures a range of products or carries out a set of functions needed by the body, this metaphor can help us to make sense of the events which are taking place.

This short session introduces some different kinds of cells, describes what they do and how they are fit for that purpose, and considers how that purpose is carried through.

Session objectives

When you have completed this session, you should be able to:

- explain in general terms how different types of cell are adapted for their particular function.

1: Fit for its purpose

Adaptation is a term used to describe how, over the course of time, the human species has come to have the characteristics which best fit its members for survival in the environment in which they live. The various parts of our bodies are adapted to perform particular functions – they are structures which are best fitted to do what they have to do.

In the same way, the cells from which our bodies are composed have become adapted to perform particular functions. The result of all these functions acting together is a living body which works more or less well, depending on how its constituent parts cope with its environment. This has similarities to the way in which the economy of a country relies upon its business and industry to cope with the demands of the wider world. So it seems reasonable to describe cells of various kinds as small businesses which contribute to the economy of the body as a whole.

The adapted cell

All cells from every plant and animal share the same basic structure. Essentially, a cell is a series of compartments which are separated from each other and from the environment outside the cell by **cell membranes**. These are structures composed of complex chemicals, which control the movement of materials around the cell and separate the areas where different activities are carried out.

The cell membranes define several kinds of structure within the cell. The most prominent is the **nucleus**. The rest of the cell is packed with other structures which all perform separate tasks.

The basic design is modified according to the main function which any particular cell performs. In the human body, cells are grouped together in **tissues** in which all the cells are the same. So groups of muscle cells make up muscle tissue and groups of insulin secreting cells make up one of the glandular tissues in the pancreas. A tissue can thus be thought of as a collection of small business concerns working together to produce a product or provide a service on a larger scale.

Although they share the same basic plan, different types of cell – such as muscle and pancreas cells – carry out very different functions and their structures and ways of working reflect that difference. Each of these types of cell is adapted to fulfil its particular role. Let's consider this in more detail.

Muscle and glandular cells

First, I would like you to draw on your own experience to think about the functions of these cells, and then we will see how they are fitted for those functions.

ACTIVITY 1 ALLOW 5 MINUTES

In this first activity, I would like you to consider muscle and pancreas cells and to identify the functions which they perform. First read the following short note about the pancreas.

The pancreas is situated close to the digestive system. It contains several different tissues, each with a function related to the use of food. One of the tissues produces insulin – the cells in this tissue are called the β (beta) cells.

Insulin is a **hormone**, a substance which is produced in one tissue and travels in the bloodstream to affect a different tissue. It controls the uptake and use of sugar by the cells. If these pancreatic cells fail to produce insulin, the affected person suffers from **diabetes mellitus**.

Now think about what you already know about muscle cells.

Finally, identify at least one function of muscle cells and pancreas β cells. Note these in the first column of *Table 1*. At the moment, you may only be able to complete this column, but by the end of this unit you should have a complete picture and be able to fill in the second column of the table.

Cell Type	Functions	Adaptations
Muscle cells		
Pancreas 'β' cells		

Table 1 The functions and adaptations of muscle cells and pancreas β cells

Commentary

From everyday experience we know that muscles bring about movements, so this is likely to be the main function that you have identified. On a cold day, we often recognise that muscle movement helps to keep us warm, so you might have added a second function – heat production.

The functions of the glandular tissue in the pancreas are less obvious to us unless we know someone with diabetes. You may have listed the production of insulin as the main function, or the control of carbohydrate metabolism as another way of expressing the same thing.

So here we have two types of cell with very different functions – movement and hormone production. Clearly, we would expect to find that each one shows its own adaptive features.

ACTIVITY 2

ALLOW 5 MINUTES

The purpose of this activity is to look at the way in which the structure of a cell is adapted for the job it does. *Figure 1* shows diagrams of a cell from leg muscle and a pancreas β cell. Study the diagrams and make a note of the structural differences which you can observe between them.

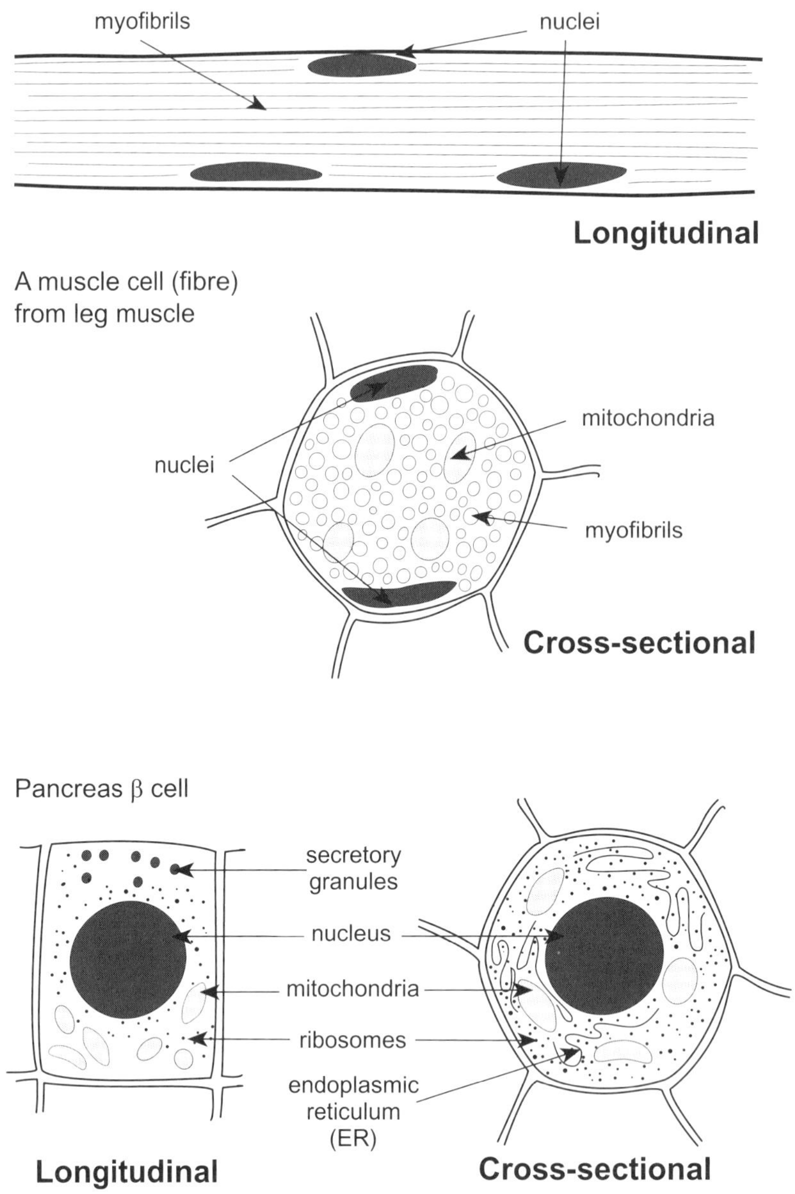

Figure 1 Muscle and pancreas cells

Commentary

The most striking difference that you may have noticed is in the relative length of the cells. Whilst the pancreatic cells are about as long as they are wide, the muscle cells are very much longer. Each one is, in fact, made up of several connected cells and is termed a muscle fibre rather than a muscle cell. A single muscle **fibre** can be as long as 30cm in a large muscle.

The next thing you may have noticed is also connected with the length. The muscle fibre has several nuclei, representing the cells from which it is composed.

The internal structures of the cells show two differences. The muscle cell is packed with **myofibrils**. These are absent from the pancreatic cell. They are the contractile elements which can shorten and lengthen, bringing about movement in the muscle.

The pancreatic cell contains a large amount of a structure labelled **endoplasmic reticulum** (ER for short). This is concerned with the manufacture of proteins. Insulin is a protein. This structure is not so prominent in the muscle cell.

Both the cells contain structures labelled **mitochondria**. We will be studying their function later in the unit.

Perhaps you would like to begin adding to the second column of *Table 1*. You can begin your list with these two adaptations:

- contractile fibres in the muscle (an adaptation for movement)
- plentiful ER in the pancreatic cell (an adaptation for making insulin).

As we progress through the unit you will look at more detailed diagrams of these two cells and you can add further adaptations as your knowledge builds.

2: Design and manufacture

If a business is to be efficient and successful, its products need to be well designed and their manufacture carefully planned. The same is also true of the cellular business, as you will see in the next activity.

ACTIVITY 3 — ALLOW 15 MINUTES

The purpose of this activity is to consider some of the similarities between the manufacturing process and the way in which a cell functions. To do this, I'd like you to use your imagination to think about the early stages of setting up a small business. You could try this as a group activity if you like.

First, select an object which you find very useful at home or at work. Choose something fairly simple but indispensable, like a thermometer or a corkscrew.

Next, imagine that this object has not yet been invented. Put yourself in the place of an inventor who sees the need for this object. You have made a drawing of what it might be like and you suspect that, if you can get it to the manufacturing stage, you will be able to set up your own business.

What are the steps that you would need to take to turn your idea into a commercial proposition? List as many as you can within the space of about 10 or 15 minutes.

Commentary

I don't know what object you chose for that activity, but it doesn't matter because the steps you'd need to take are the same. We'll stay with the example of a corkscrew to explain this.

- First, you would consult your design department as you need someone to convert your specification into a prototype corkscrew. The designer might suggest several different prototypes.
- You would have to test these prototypes, and try them out on real corks.
- Then you would choose the best one and modify it further if necessary.

Some will work well, others will fail. But hopefully, at least one will show promise and then can be further improved.

- You would keep on testing until you had an effective corkscrew at an affordable price.
- The next stage would be to figure out how to manufacture it.
- And finally, before you go into production, you would make the machinery you need in order to manufacture your corkscrew.

By now, you may be wondering what this example has to do with biochemistry. I am suggesting that these processes are similar to what has to happen before cells can begin to make something that the body needs.

Evolution as a product design

As the human species has evolved over time, various ways of meeting its needs have been 'designed' by the genetic material and tried out. Some have been successful and some have not. The activities of different cell types have been the basis for the process of **natural selection.**

Design for efficiency

In order to produce anything, a cell – just like a small business – has to make the machinery needed to manufacture the product. Like the manufacturing company, the cell needs to set up its production in the most efficient way, keeping its costs down and reducing any wasteful or harmful by-products to a minimum. And on the whole, cells are much more successful and efficient than most small businesses.

The next session considers some of the reasons for their success, starting with how the cell's machinery is made. Before that, let's summarise the main points covered so far.

Summary

- Each type of cell is adapted for a specific function – or functions.
- A tissue is a collection of cells of the same type.
- Muscle cells are long and contain contractile proteins. They are adapted for movement.
- Pancreatic cells are very different. They contain structures which are concerned with protein manufacture.
- The way in which cells are adapted is the result of a long evolutionary process of testing and redesigning their products.
- To function efficiently a cell requires efficient cellular machinery which produces few harmful or wasteful by-products.

Before you move on to Session Two, check that you have achieved the objective given at the beginning of this session and, if not, review the appropriate sections.

SESSION TWO

Cellular machine tools

Introduction

In this session, the term 'machine tools' is used to describe what is needed in order to make something to make something else.

This is exactly what the cell has to do. Before a particular cell can perform its function or manufacture its product, it has to make the necessary machinery. How it makes that machinery is the subject of this session.

The first part of this session describes how the genetic material in the nucleus contains the plans or blueprints for the machinery (that is, the proteins) and how these are sent out from the nucleus to the place where the proteins will be made. I will then describe the way in which they are made and indicate some of the things which can go wrong in the manufacturing process. The session closes with some discussion about the role of health carers in maintaining these processes.

Session objectives

When you have completed this session you should be able to:

- identify the structures on which protein synthesis takes place
- describe the mechanism of protein synthesis
- explain how protein synthesis is controlled by the nucleus
- identify ways in which health care can support this function.

1: Blueprints from the nucleus

The genetic code

In the *Genetics Unit*, we used the metaphor of a book of instructions to show how the genetic code is structured in the nucleus of every living cell. In the next activity, I would like you to revise this material if you have studied that unit or read it through if you haven't.

allow about five minutes for revision, or about 20 minutes if you haven't yet studied the Genetics Unit and this is new to you

ACTIVITY 4 ALLOW 5-20 MINUTES

This activity will help you to understand how the genetic code in the cell nucleus controls the manufacture of proteins. *Resource 2* is the section of the *Genetics Unit* dealing with the genetic code. Read it through and then answer the following questions.

1 From what is the genetic code in the nucleus made up?

2 What does the genetic code spell out?

3 What does a gene do?

Commentary

You may have noted something along these lines:

- The genetic code is contained in the structure of **deoxyribonucleic acid** (DNA). It consists of the sequence of DNA sub-units (A, T, G and C) along the long DNA strand.
- This sequence is organised into groups of sub-units (triplets) which spell out or specify the sequence of amino acids in the proteins which the cell manufactures.
- A gene is a section of the genetic code which controls the production of a particular protein or has a regulatory function.

That activity should have provided you with a clear picture of DNA as a set of instructions for making proteins. Proteins, as we will see later, carry out the activities of the living cell. DNA can be thought of as the blueprint for the production of these vital pieces of cellular machinery – a blueprint which has been tested and modified through natural selection over a very long period of time.

However, a blueprint is of no use unless it can be read and understood by the manufacturers. So now, let's take a look at how this happens in the cell.

Translating the code

The DNA bearing the instruction is inside the nucleus, and the structures which make proteins are outside. The nucleus is surrounded by a membrane boundary. We could say that the Planning Department of the small business is behind one wall, and the Manufacturing Department is behind another. How can the instructions get from the first to the second?

This is where the second type of acid mentioned in *Resource 2* comes in. **Ribonucleic acid** (RNA) is much more variable than DNA, and is found in the cell in several different forms. These forms have names which identify the job that they do. I will explain these names shortly.

It wouldn't be appropriate to spend time on the structural details of RNA and DNA here because this is only useful if you have a background in basic chemistry, and I am not assuming that everyone who studies this unit will have such knowledge. The essential feature you need to grasp is that the RNA and DNA are closely related, and DNA triplets form a pattern on which short strands of RNA can be made. These strands retain the information which was present in the genetic code, but they are small enough to get out of the nucleus, probably through pores in the surrounding membrane, and carry the information to the Manufacturing Department. Not surprisingly, the type of RNA which carries the information from the DNA in the nucleus is called **messenger RNA**.

To make this more concrete, let's focus on a particular gene. Let's think about the gene responsible for the production of adult haemoglobin (adult Hb) which was described in the case study of Karima, the child with thalassaemia in Session One of the *Genetics Unit* (see *Resource 2*). You may recall that this gene is inactive at birth because the baby has been using a special type of haemoglobin (foetal haemoglobin)which is more suitable to life in the womb. Shortly after birth, the baby needs to change over to adult Hb, so the gene which governs the production of that protein is 'switched on'.

When this happens, the section of the genetic code which constitutes that gene is copied into a strand of RNA. The DNA instructions are rewritten in RNA language. They are translated.

This piece of messenger RNA then leaves the nucleus and takes the instructions for the manufacture of adult Hb to the Manufacturing Department.

When the gene is faulty, as in Karima's case, the instructions which the messenger RNA carries are also faulty. This confuses the Manufacturing Department which will either make a defective protein instead of adult Hb, or will fail to make one entirely, sometimes with devastating consequences, as we saw when we considered the effects of thalassaemia.

The cellular Manufacturing Department is not located in one single place. It is spread throughout the area of the cell which lies outside the nucleus. We could perhaps describe it as a large number of workstations which can be moved around and grouped together as necessary. These workstations are small, spherical particles of RNA. They are known as **ribosomes**.

We'll consider how they carry out their task in a moment or two, but first let's summarise what you have found out so far in this session.

Summary

- The genetic code is based on the sequence of sub-units in DNA.
- These sub-units, in groups of three, form the triplet code.
- A gene is a section of DNA which has a particular function.
- The DNA instructions are copied into messenger RNA code.
- A faulty gene will produce an incorrect or incomplete message, so the Manufacturing Department won't be able to make the right product.
- The Manufacturing Department consists of ribosomes (or 'workstations') which are located all through the area of the cell outside the nucleus.

2: The ribosomes – workstations for protein manufacture

Let's take a closer look at these protein manufacturers.

ACTIVITY 5 ALLOW 5 MINUTES

Figure 2 shows an enlargement of the diagram of the pancreatic cell, which you looked at in *Figure 1*. This time it is shown at a higher magnification and we can see more details. After studying the diagram, describe in your own words, what you can see in this cell.

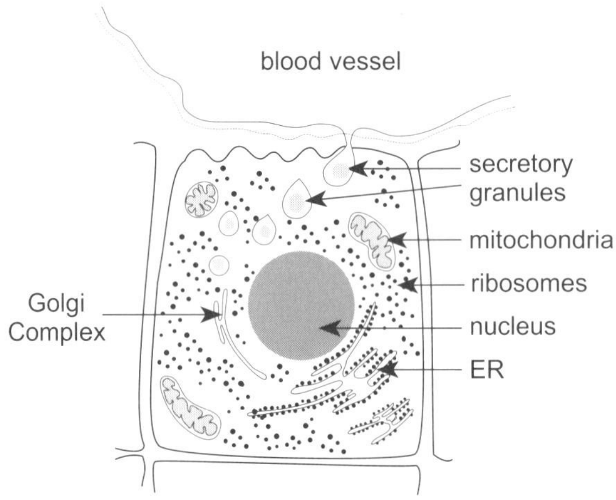

Figure 2 A β cell from the pancreas

Commentary

You may have noted down something along these lines:

'I can see that this cell is next to a small blood vessel. It contains all sorts of strangely shaped objects. I can recognise the nucleus. The other things in the cell are unfamiliar.'

This diagram shows all of the components of the cell which we will need to discuss in this unit. We will be returning to it from time to time, but for this activity we will concentrate on the function of the ribosomes.

You have probably noticed that this cell is very close to a blood vessel, which might suggest to you that it has some kind of connection with the blood. This is in fact the case, because this particular cell manufactures products which go directly out into the bloodstream for distribution.

The Manufacturing Department of the cell consists of the ribosomes. These are small spheres consisting of protein and RNA (the **ribosomal RNA**). They are put together in the nucleus and dispersed through pores in the nuclear membrane.

As you can see in the diagram, some of the ribosomes are scattered around and others are lined up along the structure labelled endoplasmic reticulum, or ER, in the lower right of the diagram.

Figure 2 is taken from a very high magnification photograph of a slice through the cell, so what you see here is what one can see when the cell has been cut across. The endoplasmic reticulum is part of an elaborate network of tubes which run through the cell. This particular section cuts through part of that system, so we can only see the shape of the cut edges where the section goes through the tubes. (Imagine cutting an apple across and finding that something has eaten into its centre – you would see a rather similar effect as you cut across the insect's track.)

When the ribosomes are first formed, they are not fixed to the ER. They are said to be 'free' ribosomes. They only become associated with the ER when they become involved in protein manufacture.

3: Assembling the raw materials

Setting up the workstations

In the nucleus of this pancreatic cell, the genes governing the manufacture of insulin are active. Messenger RNA copies of these genes are made and leave the nucleus, carrying the DNA instructions (or blueprint) for the production of this protein.

Each piece of messenger RNA meets up with and becomes attached to a cluster of free ribosomes, and the process of making insulin begins.

The cluster of ribosomes, held together by the piece of messenger RNA, becomes located on the ER. This is rather like setting up a workstation to do a particular piece of work. In this case, the task is to make a protein – insulin.

Building a protein

All proteins are made up of long chains of small building blocks – the amino acids. At this stage, you don't need to understand their chemistry, but you do need to know that the order in which the amino acids are joined together (their sequence) must be correct if the protein is to function correctly.

The messenger RNA carries instructions which, in ordinary speech, would say something like, 'Start making some insulin now. Take the first amino acid and join it to the second one, then the third, and so on, until I tell you to stop'. The workstation which has been built on the endoplasmic reticulum will obey these instructions and will join up the amino acids in the right order.

First, however, the amino acids have to be brought to the right place. Obviously some workers are needed to go and fetch the raw materials. In a healthy person, there are plenty of free amino acids inside each cell, ready to act as raw materials for making proteins. The 'workers' are made up of a third kind of RNA, called **transfer RNA**.

In fact, there are many different families of workers – one for each kind of amino acid. Their job is to link up with their specified amino acid and place it on the messenger RNA so that it can be joined to the growing protein in the correct sequence. *Figure 3* on the next page summarises this process.

Now let's briefly summarise before moving on to look at some problems of protein manufacture.

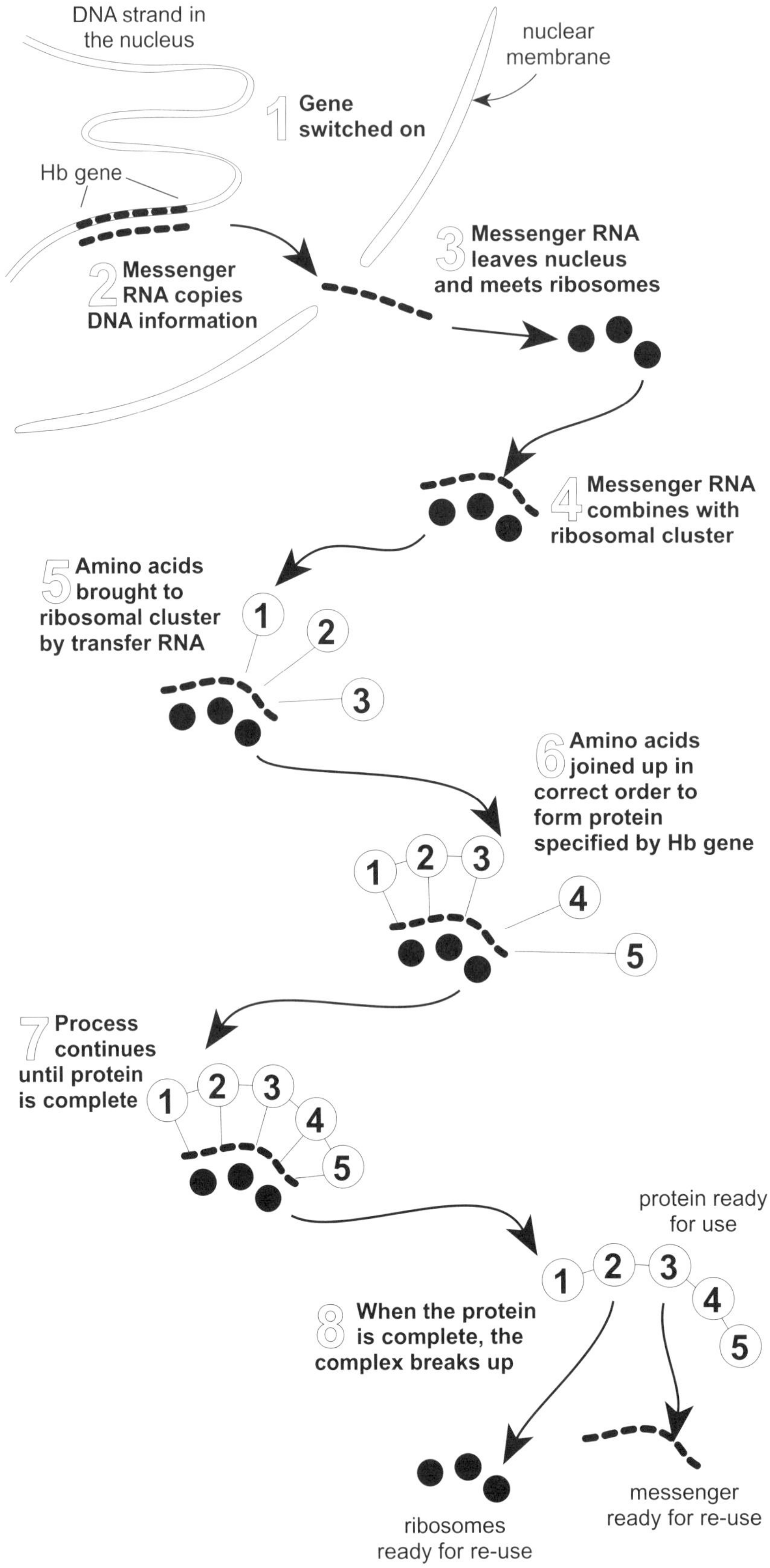

Figure 3 A summary of the events of protein synthesis within a cell

Summary

- Ribosomes consist of protein and ribosomal RNA.
- They may be free or attached to the endoplasmic reticulum.
- Messenger RNA sets up groups of ribosomes to make proteins on the ER.
- Proteins are made up of chains of amino acids.
- These must be in the correct order if the protein is to function correctly.
- The transfer 'RNA workers' bring amino acids to the ribosomal manufacturing site in the right order.

4: Problems in protein manufacture

Normally, the process I have just outlined works smoothly in the cell and proteins are made as and when needed. However, many things can go wrong.

A shortage of raw materials

If the basic materials from which proteins are built are in short supply, the production of protein will slow down. As you'll see later, the source of these building blocks is the food that we eat. Malnutrition results either from a shortage of food, or an inability to eat as a result of illness. These basic materials are then in short supply, and so the cells' ability to manufacture proteins is restricted. The consequences of this give rise to the symptoms of protein malnutrition.

ACTIVITY 6 ALLOW 5 MINUTES

The purpose of this activity is to consider the biochemical reasons for the symptoms of protein malnutrition.

Read the following case study and then consider the question at the end.

Amir is a four-year-old boy who, with his mother, has recently arrived at a feeding station in an African country devastated by drought and civil war. His mother had a second child when Amir was two, just before the war began. Amir was weaned at this age. Shortly after this, conditions in the country deteriorated and food became increasingly scarce. The animals in Amir's village all died and his family have survived on a meagre diet based on the crops they harvested and stored.

Amir weighs only as much as a healthy one-year-old baby. His skin is in a very poor condition, and his hair is wispy and reddish in colour. His arms and legs are so thin that the joints are prominent, with hardly any visible muscular tissue. In contrast, his abdomen is very swollen. The little boy is very weak, scarcely able to raise his hand to his face.

It is apparent that Amir has been on a low protein diet for some time, and is showing the symptoms of protein malnutrition. Based on what you have learned so far, can you supply a biochemical explanation for some of the symptoms described in the case study?

Commentary

You probably observed that a very striking feature of this child's condition is a failure to grow; he is very weak and lacks muscle tissue. The contractile part of the muscle is composed of protein so, if the building blocks of protein – amino acids – aren't present in the diet, muscle fibres cannot grow or be replaced. This results in restricted growth and weakness.

Amir's skin and hair are noticeably abnormal. The structural components of skin and hair are also proteins, though very different from the contractile muscle proteins. Without the right amino acids, these can't be manufactured either.

What about the abdominal swelling? This relates to a function of proteins which we haven't touched on yet, so you may not have been able to suggest an explanation. The swelling occurs because other specialised proteins are found in the fluid part of the blood (the **plasma**) where they fulfil several functions, including maintaining the normal volume of the circulating plasma. In the absence of the right amount of plasma protein, fluid accumulates in various parts of the body as the total fluid balance is disturbed. So we find that children like Amir suffer from an accumulation of fluid (**oedema**) in various areas, including the abdomen.

Amir has many other symptoms, too. His outlook is poor without an intensive feeding programme. He is not just suffering from a general shortage of food – like a person on a strict diet – but from a shortage of the essential components of protein. His condition gives us an indication of the wide range of essential functions which proteins fulfil.

In fact, the term for his condition is 'Kwashiorkor', which means ' the sickness the older child gets when the next baby is born'. It occurs when the weaned child loses the supply of protein from the mother's milk and this is not replaced by a diet containing sufficient protein of the right kind.

It is important to understand the difference between **dietary reduction**, where the total food intake is decreased but still includes protein and **dietary deficiency**, where protein is absent.

Dietary reduction causes weight loss, as in dieting and anorexia, but does not cause a breakdown in the body's own proteins unless it is severe or prolonged.

There are times when health professionals find themselves caring for people with dietary deficiency who are literally starving to death because they are unable to take in any food at all.

Under these conditions, the body uses all the reserves of fat available before starting to break down protein. The waste product from protein breakdown is the nitrogen-containing compound **urea**, which is removed from the body in the urine. As the rate of protein breakdown increases, the rate of urea output climbs. Just before death from starvation, it rises very steeply, indicating that the person has no other reserves left and is rapidly using up their own proteins.

allow three minutes, or more if you have a discussion

ACTIVITY 7

ALLOW 3 MINUTES

Drawing on your own experience and discussion with colleagues, identify and list the kinds of patients who might be described as starving to death.

You may not have encountered patients of this kind yourself, but you could use your general knowledge and awareness of current affairs to provide some examples.

Commentary

The people you may have identified will be those who are unable to eat or unable to digest the food they eat for a variety of reasons:

- people with cancer of the throat or any other severe damage to the digestive system
- people with an eating disorder, such as anorexia
- patients who are terminally ill may be unable to eat, and artificial feeding isn't always a possible option
- sometimes people survive severe brain damage but lose most of the functions of the brain, and can only be kept alive by artificial feeding. Where recovery is judged to be impossible, questions arise as to whether the feeding should continue.

In all cases where food intake has ceased the result is starvation, which will take the course we have just described.

Spanners in the works

The manufacture of protein can also be seriously affected by substances which interfere with the production of the RNA message or the joining together of the amino acids on the ribosome. Here are two examples.

- Some anti-cancer drugs affect the production of RNA. Their use relies on the fact that the cancer cells are more susceptible to them than normal cells. Therefore, the aim is to use them in concentrations which will affect the cancer cells whilst damaging normal cells as little as possible.
- All types of cells make proteins in the same way, so bacteria have ribosomes too, and several antibiotics – for example, streptomycin and chloramphenicol – directly affect bacterial ribosomes and prevent them from functioning. This is how they kill the bacteria which cause the disease. Some of these antibiotics affect human ribosomes too, but are unable to enter human cells and therefore don't normally interfere with protein manufacture there.

There are many ways in which the manufacture of protein can be upset. As you'll see in Session Three, proteins carry out most of the work inside cells. They are the machinery on which all its operations depend. So the production of proteins is fundamental to health and well-being. The symptoms of many illnesses can be traced to defects in protein manufacture. This is why maintaining a diet which contains all the right amino acids is such an essential part of individual patient care.

The importance of diet in patient care

A suitable diet must supply the cellular Manufacturing Department with all the amino acids required in sufficient quantity to produce all the proteins that the cells need.

Proteins are constructed from 22 different amino acids, some of which can be made in the body from other sources. However, about half of them cannot, and must be present in food. These are called the **essential amino acids**, and if they aren't supplied, the body will show symptoms of protein malnutrition.

As you know from *Activity* 6 and Amir's experience, one of the symptoms is muscle wastage. The myofibrils in muscle fibres are made up of two types of protein. These can slide over one another and lock into place, resulting in shortening of the length of a muscle. This is a reversible process, making muscular contraction – and therefore movement – possible. Thus, all muscular movement relies on the production of these two proteins. In turn, this relies on the availability of the right kind of amino acids.

The final activity in this session gives you an opportunity to consider some other people who may have problems with protein manufacture.

ACTIVITY 8

ALLOW **10** MINUTES

This activity will help you to relate what you have found out in this session to patient needs and patient care.

Here is a list of some 'typical' patients and clients. Select from the list those whose protein production system might be at risk in any way.

- ❑ A — A healthy mother in the third month of pregnancy
- ❑ B — A person who has suffered severe burns
- ❑ C — A 70-year-old man who has suffered a stroke and is now paralysed on his left-hand side
- ❑ D — A man of 30 who is having an operation to remove a damaged cartilage in his knee
- ❑ E — A very depressed person
- ❑ F — A young man undergoing chemotherapy for a cancer on the long bone of his leg.

Commentary

You may have come to the conclusion that nearly all these people could be experiencing some risk to their protein production system.

A The pregnant woman needs extra raw materials to supply both her needs and those of the developing baby.

B Someone who has suffered severe burns has lost protein from their damaged tissues. They will require extra protein in the diet if healing is to be successful.

C A person who has some paralysis may find eating difficult and be at risk of reduced intake of the raw materials for protein manufacture.

D A young man having a cartilage operation is not likely to be at risk, as long as he has an uneventful operation and recovery.

E A depressed person will lack the motivation to prepare and take a satisfactory diet. They may rely on cheaper convenience foods like crisps and biscuits for quick meals, and therefore have a reduced protein intake.

F Chemotherapy can interfere with protein synthesis and may also depress the appetite.

In all these cases, the provision of a well-balanced, acceptable diet will make a major contribution to well-being and recovery.

We have now taken a brief look at the way proteins are made and I have suggested that this is a fundamentally important process, since proteins do most of the work in our bodies.

The next session goes on to describe the proteins themselves and how they carry out some of this work – in other words, it describes the 'machinery' of the cell.

Summary

- Protein production can be affected by a shortage of amino acids.
- This may be caused by a lack of food or by the inability of an ill person to take an adequate diet.
- Many drugs and antibiotics interfere with protein production in humans and in bacteria which cause disease.
- Therapeutic use of these substances relies upon the fact that tumour cells and bacteria are more affected than normal human cells.
- Maintaining an adequate intake of amino acids is one way to help patients achieve a better state of health.

Before you move on to Session Three check that you have achieved the objectives given at the beginning of this session and, if not, review the appropriate sections.

SESSION THREE

The cellular machinery

Introduction

Since proteins are perhaps the most important working parts of a cell, it seems appropriate to call them the cell's machinery.

The idea of a chemical substance acting as a piece of machinery may be quite new to you, but this session aims to show you something of the amazing complexity and sophistication of these cellular machines.

The first part of the session looks at what proteins are and how one type, the enzymes, work. It then looks at the conditions they need in order to work properly and how they can sometimes go wrong, producing some of the problems we see in health care.

Session objectives

When you have completed this session you should be able to:

- describe how some proteins function as enzymes
- identify what enzymes actually do
- describe the conditions that enzymes need
- explain how they can be affected by environmental and genetic factors.

1: The nature of proteins

Fit for a purpose

All the cells in the body are derived from a single fertilised egg, and therefore all contain the same genetic information in their nuclei. Later, cells become specialised to form distinct tissues, such as muscle and pancreas.

This process of **differentiation** does not involve any loss of genetic information. It simply means that the differences between the cell types are due to different parts of the genetic code being switched on or off. In turn, this determines the kinds of proteins which are made by the cell through the processes described in the previous session.

Muscle cells make the kinds of proteins needed for their functions, and pancreatic cells make a different set of proteins which are suited to theirs. These cells are adapted for their purpose.

Any protein that is manufactured by any cell has a particular function to perform, and its structure reflects that function. It, too, is well adapted for what it has to do. Returning to our idea of the cell as a small business, we could draw a parallel with the machinery which a business needs to manufacture its products, and that needed to move things in and out of the workplace.

Consider some familiar pieces of machinery – a forklift truck, for example. Try to appreciate the complexity of its construction and the lengthy process of development, starting perhaps from the wheelbarrow, which has produced the modern, state-of-the-art equipment.

This idea can be applied to proteins as well. Let's look at an example.

One well-known protein is haemoglobin. This is the red pigment in the blood, and its job is to move oxygen around the body from where it is plentiful – the lungs – to wherever it is needed.

Like all complex chemicals, haemoglobin is made up of **atoms** combined into **molecules**. An atom is the basic unit of one of the fundamental components of all materials – an **element** such as carbon or nitrogen. A molecule is a particular combination of atoms which forms the basic unit of a recognisable material. Amino acids are small molecules which can be used to build up large ones, such as haemoglobin. So the protein haemoglobin is composed of characteristic molecules, each one constructed in the same way from amino acid components, and having the same number, type and arrangement of atoms.

We could compare each molecule to a piece of machinery – like the forklift truck – because it is constructed from particular components which allow us to recognise it and which allow it to carry out a particular job.

Taking this analogy a little further, haemoglobin does a rather similar job to a forklift truck. It picks up its load of oxygen from the store (from the lungs, where there is plenty) and carries it to wherever it is needed in the works (the cells using oxygen). However we can only take this analogy so far – it soon begins to break down.

ACTIVITY 9 ALLOW 2-3 MINUTES

This activity asks you to identify the major differences between cellular machinery and 'real' machinery.

Consider the ways that goods are moved around in a factory and oxygen is moved around the body, and then try to identify two ways in which these processes are not comparable.

1

2

Commentary

There are many differences, but the two which occurred to me are:

- In a factory, there is someone to decide what the truck should do – how much it should move and where it should take it. There is no such human agency in the cell.
- The truck has its own engines and moves about under the control of a driver. Haemoglobin molecules have no way of moving by themselves and, once again, there is no one 'in control'.

You may well have thought of other differences. The ones that I selected highlight something which is of great importance in biochemistry – the way that the actions of the cellular machinery are controlled and regulated. We will return to this in Session 7.

Another point you might have mentioned is the development aspect of forklift truck construction. I said earlier that this had taken place over a long period, beginning with our first attempts at using machines to move things around. You might have questioned the extent to which the analogy stands up here. It is very relevant to ask this kind of question, and the next activity will help to provide some answers.

ACTIVITY 10

ALLOW 5 MINUTES

First, read the following short account of our current understanding of haemoglobin manufacture. When you've read it, try answering the question which follows.

The structure of haemoglobin is very similar in all animals with backbones (vertebrates). Less advanced animals have other, less efficient oxygen-carrying molecules in their blood.

The group of vertebrates to which we belong, the mammals, contains our closest relatives, the great apes. Haemoglobins from humans and chimpanzees are exactly the same, and there is only one different amino acid between human and gorilla haemoglobin. Monkeys, who are less closely related to us, have haemoglobin which differs from human Hb in about eight amino acids.

The type of haemoglobin manufactured by the early human foetus is different from that made by the baby, and different again from the adult type. Now that the structure of our chromosomes is being worked out, the position of the genes for these different haemoglobins is known. It turns out that there are five of these, close together. They are switched on in the same order that they occur on the chromosome.

Figure 4 below shows the human haemoglobin genes on chromosome 41. The 'p' or pseudogene is thought to be a 'design failure'; it doesn't work but has somehow persisted. Adult haemoglobin is mostly composed of the 'A' type, with a little (two per cent) of 'a'.

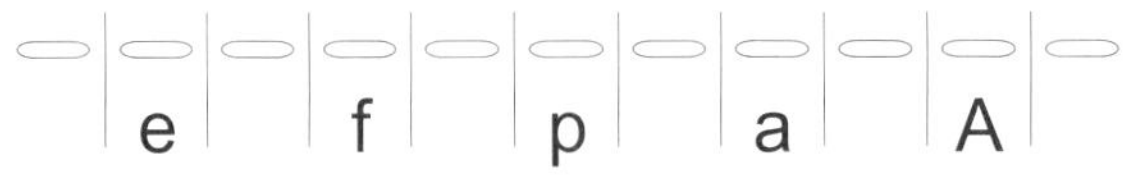

e = very early embryo
f = foetus
p = 'useless' gene—it doesn't produce anything
a = adult (minority)
A = adult (majority)

Figure 4 Human haemoglobin genes on chromosome 41

Now that you have this information, do you think that there is any evidence of a development process for haemoglobin, like the development process of a machine?

Commentary

On the basis of this evidence, you may agree that it looks as if the animals who are more advanced, including humans, have developed a more successful way of moving oxygen around their bodies.

The fact that there are three different types of haemoglobin in humans shows that developments have occurred, and some have proved to be more successful than others under different conditions in the embryo, the foetus, and the adult. The non-workable pseudogene is an example of a dead end in this process.

From this, we might conclude that human haemoglobin represents the present 'state of the art' in oxygen transport machinery. This is true for all the proteins which we are going to discuss – they all represent the best possible pieces of machinery for their particular jobs.

Summary

- The structure of any protein reflects its function in the cell.
- Specialised cells contain specialised proteins.
- Protein structures have developed through time as, through natural selection, they have become increasingly well adapted for their particular tasks.

Now we need to consider some aspects of the structure of proteins.

Protein structure

All protein molecules are built up from amino acids, as described in Session Two. There are about 20 different kinds of amino acid available for use in living cells, and each type consists of relatively small molecules which are the basic building blocks from which the protein machinery is constructed.

The smallest of the amino acids consists of a molecule of ten atoms. The others range upwards in size to about 30 atoms.

All amino acid molecules contain nitrogen atoms. They have other similar features too, allowing them to be joined together in long chains through the same kind of linkages. However, there are also differences between them, so they behave in different ways. When they are joined up into a protein molecule, its character will be determined by the behaviour of all its amino acid components.

Proteins are mostly very large molecules consisting of hundreds, or even thousands, of amino acids linked together in chains. Insulin is the smallest that we will consider in this unit, consisting of 51 amino acids. Haemoglobin is much larger, with nearly 600 amino acids, and collagen (a structural protein found in skin) has about 3,000 amino acids per molecule.

The fact that amino acids can be joined up to form very long chains is of great importance to the cell. Let's explore one of the reasons for this.

ACTIVITY 11

ALLOW 5 MINUTES

Imagine that you are going to build a machine from a set of components that can all be fitted together in any possible combination. You have a plentiful supply of four kinds of components – A, B, C and D – and the machine you are building is small, consisting of only four components. Work out how many different machines you could make with the components you have to hand.

Commentary

If you don't have a mathematical education, you probably tackled this activity from first principles.

You could build four different machines using the same components each time – AAAA, BBBB and so on. Then for each of these, you could change one letter – AAAB, AAAC and so on. Next, you could make a second substitution – AABB, AABC.

This process could last quite a time and you probably got tired of it before you exhausted all the possibilities.

Another way of tackling this would be to use some simple maths. We have four possible choices for the first letter, and four choices for the second letter of each of these.

So we have 4 x 4 = 16 choices for the first two letters. Now what about the third and fourth?

ACTIVITY 12

ALLOW 10 MINUTES

1. Using the information above, work out the total number of different four-component machines you could build with the four types of components used in *Activity 11*.

2. How do you think we could work out the number of possible insulin-sized proteins which cells could manufacture from the 20 different available amino acids? I don't want you to work this one out; just describe how it could be done.

Commentary

1. Continuing the process we started in *Activity 11* gives an answer of:

4 x 4 x 4 x 4 = 256

So we could build a large number of different machines by combining the four components together in different ways.

2. Using the same system to work out how many different proteins of about 50 amino acids we could make from a supply of 20 different amino acids, we'd need to multiply 20 by itself 50 times – which is a very large number.

What this exercise shows is that, because amino acids come in 20 varieties and can all be joined together in any combination, there is an infinitely large number of possibilities for building protein molecules, giving us a tremendous variety of proteins.

It would be a mistake to think of proteins simply as varied chains of amino acids. When the large protein molecules are formed on the ribosomes, the chains become folded and coiled to give elaborate structures.

In addition, a protein molecule may consist of two or more different chains and may include a non-protein component.

Haemoglobin shows all three features I have just described. Adult human Hb consists of four amino acid chains (two identical α chains and two identical β chains). The chains are linked together and coiled in a very precise way.

Figure 5 is a very simplified diagram of a human adult Hb molecule.

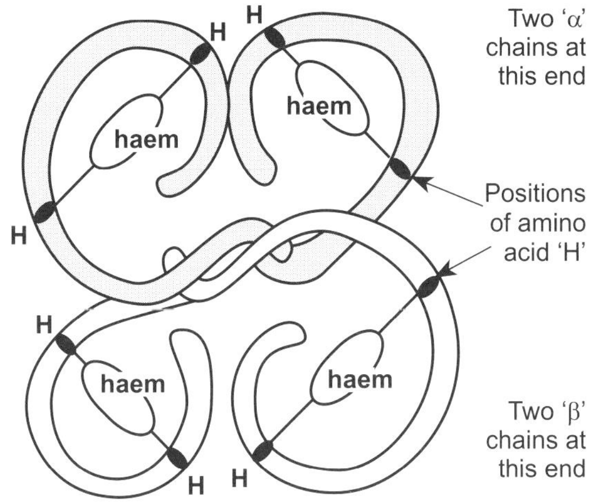

Figure 5 Diagrammatic representation of a haemoglobin molecule (adult Hb)

The folds and coils of this molecule are not just random shapes. They are essential if haemoglobin is to function efficiently in its capacity as an oxygen carrier.

You may notice that the molecule contains four spaces into which a non-protein substance **haem** fits neatly. It needs two special linkage points in order to fit. These are provided by one particular amino acid, which is shown as 'amino acid H' on the diagram. The way that the chains are coiled puts two Hs in just the right places relative to one another on each chain. This allows the haemoglobin molecule as a whole to work very efficiently in transporting oxygen.

This description of haemoglobin may have helped to give you an idea of how varied proteins can be. There are about 10,000 different ones in the human body and each one has a complex shape enabling it to perform a particular function.

Many proteins (for example, haemoglobin and the plasma proteins) have molecules with rounded shapes. Others are long and thin (for example, the contractile proteins in muscle). Others again form flat flexible sheets (for example **keratin**, the protein which forms an important structural element in skin). In all these cases, the shape fits the protein for its role. This feature is seen to greatest extent in the proteins which do most of the work in the cell – the **enzymes**. This important group of proteins is described in the next section.

Summary

- Because proteins are made up from 20 different amino acids, their variety is very great.
- The variety arises from the order of the amino acids in the protein chains, the way the chains are folded and combined, and the additional non-protein components which they may contain.
- The structure of a protein must be exactly right if it is to function properly.

2: Enzymes

This section describes the unique properties of enzymes and the way that they operate. We'll look at the example of a particular enzyme (**catalase**) as an illustration of the way that enzymes have particular functions in the cell. Let's begin by considering what the term enzyme actually means.

Enzymes as cellular machinery

The sum of all the chemical changes which take place in cells is termed **metabolic activity**. This is really the 'business' that the cell undertakes, and consists of taking in raw materials, breaking them down into their components, and rebuilding these into whatever the cell needs to make for itself or to export elsewhere.

Many of these chemical changes can be brought about in chemical laboratories, using the standard techniques of the chemist. However, in the test tube, these reactions usually require heat or additional chemicals to make them happen. The wonderful thing about biochemistry is that all these changes can be brought about without the need for any such conditions. They happen at body temperature and in conditions compatible with life – **physiological** conditions. This is possible because of the presence of proteins whose structure allows them to act as enzymes, biological agents which encourage metabolic reactions to take place under the conditions in the cell.

All enzymes are proteins, but not all proteins are enzymes. Haemoglobin is not an enzyme, but its structure has features in common with enzymes, and will be a useful point of reference. You have already seen how the haemoglobin molecule is folded in a particular way which produces spaces and combining sites for haem. Enzymes are similar in that their structures provide spaces and combining sites for the materials on which they are going to work.

Since these sites are just the right shape to fit particular materials, it follows that they will not fit many others. So most enzymes will only work on a few related materials. They exhibit specificity.

Enzymes are named according to the work that they do. All their names end in '-ase', and the rest of the name indicates the type of job and the material on which the enzyme works. All enzymes have a systematic name of this kind which arises from the classification system we have just described. Many also have another name which is more commonly used, as it is the name they were given when first discovered, before the international classification system was set up.

ACTIVITY 13

The purpose of this activity is to introduce some enzymes which can be used to help in diagnosis.

Laboratory request forms often ask for the level of certain enzymes in a patient's blood. For example, a person admitted with chest pain may be suspected of having a heart attack (myocardial infarction). It is usual to investigate the levels of some enzymes in such a person to confirm the diagnosis.

See if you can take the opportunity to find out what these enzymes are and why their levels in the blood are significant.

You could discuss this with an experienced general nurse, a doctor, or a member of the laboratory staff.

Commentary

The two enzymes which are most often measured are AST and LDH.

LDH stands for lactate dehydrogenase, an enzyme which we will meet later in this unit. AST stands for aspartate transaminase. Both these enzymes are present in heart muscle cells. When the heart muscle fibres are damaged, their contents are released into the blood. The levels of these enzymes rise about 12 hours after damage occurs, and remain high as the dead tissue in the heart is gradually destroyed over a period of about a week.

How do they do it?

All the chemical changes which enzymes bring about would only happen very, very slowly, if at all, in the absence of enzymes.

An enzyme speeds up a chemical change (a **reaction**) so that something which could only happen infinitely slowly takes place in milliseconds. Any substance which speeds up reactions like this is a **catalyst**, so enzymes are biological catalysts.

They speed up reactions by combining with the material on which they work (their **substrate**). The substrate fits into a particular place in the enzyme structure, in the same way as haem in haemoglobin. This causes stresses within the substrate molecule which make it much more ready to undergo a reaction.

Let's make this more concrete by considering a very widely distributed enzyme. Its popular name is **catalase**, and its job is to destroy hydrogen peroxide.

You might think that this enzyme would not be terribly useful in living cells, because we don't normally think of them as containing hydrogen peroxide. However, we shall see that it does serve a very useful function.

ACTIVITY 14 — ALLOW 15 MINUTES

This activity will enable you to identify the effects of catalase. To do this, I'd like you to try out a little chemistry. You will need to buy some 10 volume hydrogen peroxide from the chemist – you only need an egg cup full for this experiment.

Place a few drops on a saucer and see if any small bubbles form as it stands. Jot down your observations.

Then place a small piece of living material on the solution and observe the result. You could use a drop of your own saliva, as this contains living cells which have come loose from the lining of your mouth; or a piece of freshly cut, uncooked plant material, such as potato or apple; or a tiny piece of fresh meat.

You could try a number of different substances.

Record your observations each time.

ACTIVITY 14 *continued*

Finally, try small pieces of cooked plant or animal material and record your observations again.

Commentary

Did you notice a few small bubbles in the drop of hydrogen peroxide solution before you added anything to it? These were bubbles of oxygen. Each hydrogen peroxide molecule consists of two atoms of oxygen and two of hydrogen. They break down very slowly to release the oxygen which, being a gas, forms these bubbles as it escapes.

If you added anything with living plant or animal cells to this solution, you would expect to see a much more rapid rate of oxygen release, shown by a rapid stream of bubbles – even a froth. Something has increased the speed at which the hydrogen peroxide molecules are breaking down.

That something is the enzyme catalase. You'll probably have been able to detect it in nearly all the substances you tried during that short experiment, as long as they haven't been cooked.

You would certainly find it in all really fresh animal tissues, but not in cooked material.

You might wonder why catalase is so widespread and what use it can be in living tissues. In fact, hydrogen peroxide is produced in all tissues at the end of a series of reactions which protect cells from the harmful effects of oxygen.

We tend to think of oxygen as vital for life – as indeed it is – but our use of oxygen carries a penalty. In its ordinary form as a gas, oxygen atoms go round in pairs and are totally harmless. However, under some circumstances where oxygen is plentiful, small amounts of an alternative form of the oxygen molecule are produced. These are called **free radicals**. They are highly reactive and do a lot of damage inside cells if they are not immediately removed. They are formed as a by-product of normal metabolic reactions which involve oxygen.

Free radicals are destroyed by a series of enzymes specifically structured to deal with them. An intermediate stage in this chain of events is the production of hydrogen peroxide. Catalase, which destroys this substance and frees the harmless 'ordinary' oxygen, is the last stage in this protective system. We can summarise this as:

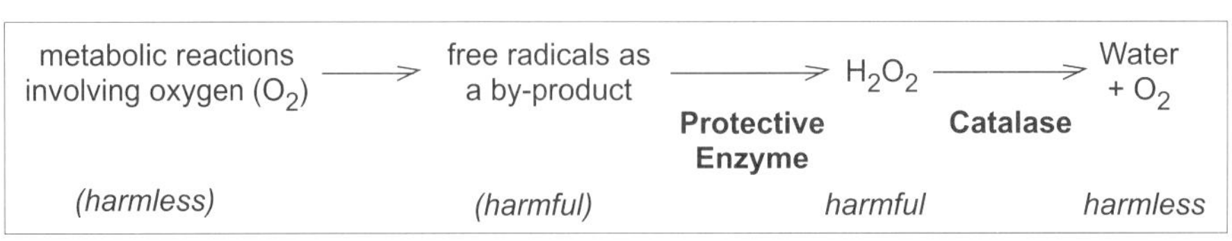

Notice the convention used in biochemistry of showing a reaction by means of an arrow and writing the name of the enzyme which brings it about immediately below.

The harmful effects of free radicals are shown by the serious symptoms of poisoning with the weedkiller paraquat. This substance produces a large amount of free radicals, more than the protective enzymes can cope with. The free radicals attack vital parts of the cell and start a chain reaction, which produces more and more free radicals, with fatal results.

Catalase speeds up the breakdown of hydrogen peroxide by about a million times. The enzyme molecule has four combining sites into which hydrogen peroxide, and nothing else, can fit very neatly. This distorts the hydrogen peroxide molecules just enough for them to decompose, giving off the oxygen gas and leaving just water behind. As soon as this has happened, the combining site is free to take in another molecule of hydrogen peroxide.

This example demonstrates three important features of the way enzymes work in general.

- It is specific, accepting only its substrate, hydrogen peroxide.
- It speeds up the rate at which a reaction which would normally be extremely slow takes place. Working flat out, each catalase molecule can deal with nearly 100,000 molecules of hydrogen peroxide per second.
- Although it combines with its substrate, it remains essentially unchanged itself. When the oxygen is liberated, the enzyme is ready to deal with the next peroxide molecule. It is therefore acting as a catalyst.

These properties of catalase have led to the use of hydrogen peroxide for some clinical purposes.

ACTIVITY 15

This activity gives you the opportunity to investigate the use of hydrogen peroxide as a wound cleanser.

Some people recommend the use of dilute hydrogen peroxide as a cleanser for deep wounds infected with bacteria which can only thrive in an oxygen-free environment – that is, anaerobic bacteria. Other people contend that this practice is now out of date and that hydrogen peroxide should not be used in this way.

Discuss with your colleagues the benefits and disadvantages of this wound cleanser, using the understanding you have already gained in this unit. Identify one reason in favour of its use, and one against.

Hydrogen peroxide is useful as a wound cleanser because –

Its disadvantages are that –

Commentary

The rationale for the use of hydrogen peroxide in wound care was that catalase in the exposed tissues would react with it to release oxygen, which would kill off the anaerobic bacteria.

The case against its use is that the large amounts of oxygen produced would lead to free radical production and tissue damage.

In practice, hydrogen peroxide was found to be effective as a cleanser for malodorous wounds that were difficult to treat in other ways. Those who used it were careful to prevent it from coming into contact with the surrounding skin, as they recognised the damage it caused.

However, nowadays it is possible to use an antibiotic (metranidazole) to remove the anaerobic bacteria from such wounds. This destroys the source of the offensive smell, so the use of such a potentially damaging wound cleanser is no longer necessary.

This example shows how accepted practice in health care can change over time, and that an understanding of the biochemistry of cells can be used to inform practice in wound care.

Resource 3 provides some useful background reading about the treatment of one type of difficult wound. You might like to read it at this point.

Summary

- Enzymes bring about all metabolic activity.
- They enable reactions which would normally be infinitely slow to occur rapidly under physiological conditions. They act as biological catalysts.
- Each enzyme is specific for a single substrate or for a few closely related ones.
- Catalase is a widely distributed enzyme. It serves a protective function by removing hydrogen peroxide wherever it is formed.

Ideal operating conditions

Although enzymes act as catalysts, they do undergo some changes as they carry out their role, so it's not strictly correct to say that they are unchanged in the process. They suffer wear and tear as they are exposed to their working conditions.

While there are no health and safety regulations in the small business we are considering, the correct working environment is very important. Enzymes are affected by many aspects of their environment, and here we would like to consider four of them.

1 temperature	3 presence of other substances
2 acidity	4 pressure

The first is the temperature of the surroundings. When the temperature is high, the molecules of any substance move around faster, because the heat applied has been converted to movement energy in those molecules. Thinking about the enzyme catalase again, we could picture the molecules of its substrate (hydrogen peroxide) moving around faster and faster. What would the result be?

ACTIVITY 16 ALLOW 2 MINUTES

Imagine that you are in a crowded room with lots of other people, dancing to your favourite music. As the tempo increases and the dancers move more quickly, what happens to your contacts with the other dancers?

Commentary

I don't suppose you needed to reflect long on this one. The number of collisions between dancers will increase as they move around more quickly.

This is not a trivial example, because it mirrors the situation between enzyme and substrate. The enzyme can only act when it collides with a substrate molecule at random. There is no force attracting one to the other.

In a liquid medium, like the inside of a cell, freely-moving molecules of all kinds are constantly colliding with each other. As the temperature increases, their speed of movement increases too, and the rate of collisions rises; so as the temperature rises, the enzymes work faster.

However, after a certain point, the increased temperature starts to affect the atoms in the enzyme too. They also move and the structure of the enzyme molecule becomes disrupted. It can no longer combine with its substrate, so its rate of work goes down.

There is therefore a balance point where the enzyme is working as fast as it can without being damaged. This is called the **optimum temperature** for that enzyme. As you might expect, the optimum temperature for human enzymes is normal body temperature. They are stable at a lower temperature, but become inactivated if they are kept at higher temperatures for any length of time.

The second environmental factor which affects enzyme activity is the degree of acidity of the surrounding liquid. This is measured on the pH scale, and the measurement tells us the level of concentration of free hydrogen atoms in the liquid. Exactly why hydrogen makes something acidic doesn't concern us here, since we want to avoid unnecessary chemistry. But you do need to be aware of what the pH scale means.

Measurements on the pH scale extend from 0 to 14. The middle of the scale (pH 7) represents a situation which is neither acidic nor alkaline; it is neutral. Lower values on the scale are increasingly acid and higher values increasingly alkaline.

Normally, the environment inside the cell is close to neutral, and this is the optimum pH for the activity of most human enzymes. They work best under these conditions and are inactivated if the pH rises or falls too far.

Thirdly, the presence or absence of other chemicals in the environment affects the activity of enzymes. Some substances are essential if an enzyme is to work properly. They help with its activity, taking part in its interaction with the substrate. Many of these are substances needed in the body in very small amounts, such as vitamins and essential minerals.

Other substances interfere with enzyme activity. These are called **inhibitors.** Poisons often exert their effect by inhibiting vital enzymes. Cyanide, for example, kills by completely inhibiting one of the enzymes taking part in the use of oxygen from the air, so it effectively deprives the body of oxygen.

Finally, a group of factors which we termed 'pressure' affect enzyme activity. The large enzyme molecules are not very stable, and their complex structures can be disturbed if they are subject to mechanical pressure, such as that resulting from vigorous stirring or shaking. This has practical implications, as the next activity shows.

ACTIVITY 17

ALLOW 5 MINUTES

Wounds containing dead tissue (**necrotic tissue**) will not heal unless this tissue is removed. The removal of such material is termed **debridement** of the wound.

Resource 4 contains information from three sources about the use of a debriding agent which contains bacterial enzymes. These enzymes break up and liquify dead tissue, making it easy to remove from the wound without further trauma to the injured person. Please read *Resource 4* and find the answers to the following questions.

1 What are the names of the enzymes contained in this preparation?

2 What two precautions are recommended to ensure that the enzymes remain active?

3 Can you explain the need for these precautions?

Commentary

Reading through *Resource 4*, you will have discovered that:

1. The two enzymes are **streptokinase** and **streptodornase** – like all enzymes, their names end in '-ase'.
2. To ensure that these enzymes remain active, they need cold storage and gentle treatment.
3. The reasons for these precautions are that both heat and shaking disturb the complex structure of the enzymes, thus rendering them inactive.

From this brief description, I hope it is clear that the operating conditions inside each cell have to be carefully controlled if the cellular machinery is to work properly. Despite the presence of a number of control systems, which we will consider later on, the working conditions of the enzymes can still be upset with serious consequences. We will now look at some of the factors which disturb or damage the cellular machinery.

Summary

- Each enzyme works best under a certain set of conditions.
- Its rate of activity is governed by the temperature, pH and pressure of its surroundings as well as the presence of other substances.
- An enzyme molecule acts when it collides with its substrate in the right way. This is a random process.
- The internal conditions of the cell need to be carefully controlled if the enzymes are to function correctly.

3: Faulty machinery

Environmental change

Problems can arise in the cellular machinery if the operating conditions are disturbed. We have already mentioned the effect of a poison. This is just like throwing a spanner into the works, as it jams the whole system.

Less dramatic effects happen if the other conditions change more gradually, as the next activity indicates.

ACTIVITY 18 ALLOW 10 MINUTES

The purpose of this activity is to identify what is happening to the body's cellular machinery when we experience a rise in body temperature.

In health care, we often encounter people with a raised body temperature – people who are **pyrexial**. You have probably had this experience yourself.

It happens because the normal mechanisms for controlling our internal temperature break down, or because our internal temperature regulator is reset. Think about an occasion when you were pyrexial, and write a short paragraph about your experience. What was the cause? What did it feel like?

Using the information you have gained from this unit, see if you can give a simple explanation of what was going on inside your body during the feverish episode.

Commentary

Most people have been pyrexial at some time in their life as a result of an infection. Disease-causing bacteria or viruses have entered the body and started to multiply. Most of us know how ill we feel with **influenza** – a relatively mild illness in otherwise healthy people. We feel shivery, sick, have a headache and pains all over. Some of us will have experienced more serious respiratory illnesses which can follow, in which the pyrexial episodes go on for a longer time, leading to **pneumonia**. The pyrexia takes away the appetite and makes the patient feel very weak and ill.

Many young people suffer from **glandular fever**, in which a fluctuating high temperature can persist for months. This leaves people feeling very weak, depressed and generally out of sorts.

So nearly all of us can describe the physical symptoms of pyrexia and the extreme weakness and debility which is the result of repeated or prolonged pyrexial episodes.

Why should this happen? What use is a high temperature?

You might have suggested that the pyrexia was a useful response to infection, since it is one way of damaging the infecting bacteria by upsetting their cellular machinery. Of course, over a long period, it disturbs our own cells as well, and this is part of the reason for the discomfort which occurs and the need to remove toxic by-products and rebuild cellular structures during convalescence.

There are other ways in which faults develop in the cellular machinery. The symptoms of many illnesses are caused by the effect of the illness on a crucial enzyme. For example, liver failure can arise from a number of causes, but all of them affect the liver enzymes which deal with the waste products from protein breakdown. These poisonous waste products will then accumulate, leading to severe symptoms and eventual death.

Genetic damage

Finally – and very importantly – we need to consider a group of conditions arising from genetic causes, which result in enzyme defects from the time of birth. These conditions are known as **inborn errors of metabolism**. Some of them are well known, such as **phenylketonuria** (PKU).

ACTIVITY 19 — ALLOW 15 MINUTES

Using a case study of a family affected by PKU, this activity highlights the health-care implications of an enzyme defect arising from genetic causes.

At the end of the case study are a number of questions which you could answer yourself, or which could form the subject for a group discussion.

Maura and Fergal are a young couple from Eire. Maura is 23, and has spent some of her life in a unit for people with learning difficulties. She was able to develop the skills of independent living and, after a time, moved out into supervised accommodation. At the age of 20, she decided that she wanted to be more independent and moved to Liverpool with her much older sister, Maeve, who has a job and a family there.

Maura helps care for her sister's children and has a part-time job in a factory, where she met Fergal, with whom she now lives.

They are thinking of starting a family, but Maura's sister is concerned as she remembers someone telling her that Maura should be sterilised because she should never have children. Maeve was away from home when Maura was a baby, but she can vaguely recall that Maura had to have a special diet until she was about 15. The sisters have now lost touch with their family and with the unit where Maura lived.

After talking to members of the primary health care team, Fergal decides that they must discover what the problem might be. He finds the address of the unit where Maura lived, and their present GP writes for information. It turns out that Maura was diagnosed at birth as having PKU. This was detected by the Guthrie test, which detects any excess of a particular amino acid in the blood. The illness is caused by an enzyme defect which prevents this amino acid from being properly used in the body. It accumulates and causes damage to the developing brain, especially in the early months of life.

Maura was born in 1969. She was fortunate, because before the 1960s this condition was undetectable until the damage had been done, so most affected babies became severely brain damaged. The treatment which Maura received was very difficult for her and her family. She had to eat an entirely artificial diet from which all the amino acid in question had been removed. She can remember this well, and also recalls that as soon as she was away from her mother's supervision, she started to eat other foods and finally reverted to a normal diet, although her mother said she had to stick to her diet for life.

The treatment was obviously a success for Maura and protected her developing brain from the effects of this amino acid. She has only mild learning difficulties and is a strong, healthy woman with a very pleasant personality. She tends to become agitated in unfamiliar situations, but copes well in her supportive family.

The primary health care team had not seen an adult with PKU until last year, when they found they had two patients with this condition. Fortunately for Maura, because of this they had located a source of specialist advice and were able to refer Maura and Fergal to a clinic where appropriate tests and counselling were available.

When Maura's blood was tested, it was found to contain higher than normal levels of the amino acid. She and Fergal were told that, if Maura becomes pregnant, there is a high probability that the baby will be affected by this amino acid and that its brain and the development of its head will be retarded in the womb. The clinic staff stressed that the baby's disabilities are likely to be much more severe than Maura's.

However, they also say that if Maura goes back on the artificial diet before becoming pregnant and sticks to it right through the pregnancy, she has a good chance of producing a normal baby – though they obviously can't give any guarantees.

Fergal, Maura and Maeve have a lot to think about.

1 Using the information in the case study, write a short explanation of Maura's present condition.

ACTIVITY 19 *continued*

2 Can you suggest why she was given the advice about her diet at the specialist clinic?

3 What problems do you think Maura and Fergal might face if they decide to go ahead and start a family?

4 This condition is particularly common in Ireland. Can you suggest any ethical problems facing health carers in a country with a high incidence of treated PKU?

Commentary

1 As we know, Maura has a genetic defect which makes it impossible for her body to use a particular amino acid. She was given a diet from which this amino acid was eliminated. It seems she stuck to this pretty well in her early years, because she has avoided severe brain damage. It is impossible to say if her mild learning difficulties are due to lapses from the diet or if they are entirely separate from her PKU.

2 If Maura gets back on the diet, she won't have such high levels of this amino acid in her blood. If she becomes pregnant and her baby also has PKU, it won't be affected in the womb by the amino acids in Maura's blood.

3 Maura will have to endure the diet throughout her pregnancy. If her baby does have PKU, it will have to be fed this diet and carefully monitored during childhood. This can be very difficult for parents.

4 Since children have been treated by diet for this disease, more affected children are surviving into their reproductive years. If an affected woman becomes pregnant and her unborn baby has PKU, the baby's future outlook is poor. This raises ethical questions about the terminations of such pregnancies.

On a wider scale, however, there are questions about screening to identify potential mothers who have PKU but are not aware of the fact. If such women are detected, how can they be encouraged to stick to the artificial diet during pregnancy? These are rather difficult questions concerning the relative rights of the mother and her unborn child.

Specialist paediatric nurses have an important role in communities with a high incidence of PKU. They are responsible for many aspects of the care of children and their families, including screening, monitoring and counselling. They need an in-depth knowledge of all aspects of this condition, including its biochemistry.

The example of PKU shows how the effects of a fault in the enzymatic machinery can be very widespread. There are many other examples of this kind of inherited disorder, including **porphyria** – the disease thought to have been responsible for the 'madness' of King George III in the late eighteenth century, and the more common familial **hypocholesterolaemia.** The latter is considered to be a cause of heart attacks in people under 60 years of age. It is characterised by high levels of cholesterol, a fatty substance, in the blood. These arise because there is a defect in one of the enzymes responsible for the uptake and use of cholesterol in the cells. Thus more cholesterol than usual remains in the blood, and it can build up on the walls of blood vessels, eventually blocking the supply to the heart.

Unusual drug reactions

When a person is given a drug, the cellular machinery deals with it in some way or other. The body metabolises the drug, just as it metabolises any food material. Usually this happens by a process of breaking down the drug or by combining it with something else to render it harmless to the cells. These processes are brought about by enzymes, so enzymes are crucial to the body's response to drugs.

All drugs are carefully tested to ensure that they can be easily metabolised by a 'normal' set of enzymes. However, a minority of people with no apparent symptoms have defects in one or more enzymes. These are the people who show an abnormal drug reaction because their cellular machinery cannot cope with a particular drug.

There is a group of people who are called **slow metabolisers**, because they have various genetically-determined enzyme defects which prevent them from dealing with common drugs in the normal way. One of the earliest such defects to be discovered related to the drug **isoniazid**, used to treat tuberculosis. Some patients reacted severely to the normal dose of this drug. Because it persisted in their bodies, they required much lower doses than normal to produce a therapeutic effect. This was due to a defect in one of the enzymes responsible for the removal of isoniazid from the bloodstream.

There are many other examples of conditions in which drugs are only slowly removed from the body. It used to be considered that these represented unusual or 'idiosyncratic' responses, but now understanding is growing that the variety of drug response is much greater, and that its basis is often genetic. It is important for nurses and all health carers to be aware of the individual nature of a person's response to the drugs that are given as part of their treatment.

Summary

- Cellular enzyme activity can be upset during an illness when the conditions in the cell are disturbed.

- Some illnesses are caused by a genetic defect which leads to a malfunctioning enzyme, such as PKU, porphyria and familial hypercholesterolaemia.
- Certain responses to drugs are caused by unsuspected defects in the enzymes which metabolise these drugs.

This concludes our investigation of enzymes and their role in the body. I have only mentioned a few of the hundreds of enzymes which the body is known to produce. Nevertheless, I hope that this has given you an insight into their nature, the way they work and how important it is to maintain their integrity.

Before you move on to Session Four, check that you have achieved the objectives given at the beginning of this session and, if not, review the appropriate sections.

SESSION FOUR

Importing raw materials

Introduction

The first three sessions have described how the cellular small business prepares to go into production. It has the blueprints, it constructs the machine tools, and then it sets up its cellular machinery. The next stage, then, is to import the raw materials.

This session describes some of these raw materials – what they are, where they come from, and how they are brought to the cell. The last part of the session considers the very important question of how these raw materials actually get through the factory gates – in other words, into the cell itself.

Session objectives

When you have completed this session, you should be able to:

- describe how food materials are broken down into a relatively small number of basic materials
- explain how theses materials – especially glucose – get into the cell
- describe how this range of materials can be used to make all the cell's products.

1: Food as raw materials

I always feel rather saddened when students in the health care professions say that, 'Biochemistry is so boring'. When pressed, they often complain about having to learn the constituents of food and the formulae for sugars. This seems to be something that demotivates a lot of people.

I feel sad because my own experience of this aspect of the subject was different and very positive – a revelation, in fact. I can remember first hearing about the building blocks of food and the way they can be taken apart and rebuilt. Suddenly, it seemed that I could see a pattern underlying all living processes, and this struck me as a very wonderful thing.

In this session, I will try to convey some of this pattern and order, and avoid the approach which results in the response just described. So we will be focusing on broad principles rather than details.

The food we eat provides all the raw materials for our cellular machinery to work on. Most of what we eat has been produced by plants and animals, from simple substances extracted from the earth, the air and the sea.

Plants are the basic providers. They have the unique ability to convert the small molecules of carbon dioxide, water and nitrogen into very large molecules such as proteins.

Animals eat plants. They break down large plant molecules into their components and rebuild them into animal proteins and other large molecules. Humans eat both plants and animals, and carry out exactly the same process.

The food we eat is mainly composed of substances which are classified into three groups:

- carbohydrates
- proteins
- fats.

Carbohydrates and proteins consist of molecules of a particular type, called **macromolecules**. These are very large molecules built up from smaller units which can be joined together to produce a vast range of different structures. In Session Two we encountered another group of macromolecules, the nucleic acids. In Sessions Two and Three we also saw how proteins are constructed. Let's revise the ideas from these previous sessions to see how nucleic acids and proteins conform to this description of macromolecules.

ACTIVITY 20 ALLOW 5 MINUTES

Look back through Sessions Two and Three to remind yourself of how nucleic acids and proteins are constructed.

Nucleic acids are built up from the four DNA sub-units __ , __ , __ and __.

Proteins are built up from __________ ________ .

Commentary

Nucleic acids are made up of large numbers of their basic sub-units, A,T,G and C. No further details of their structure are required in this introductory course. The important thing to remember is that the tremendous variety of the nucleic acids is responsible for much of the variation between people, and it is all due to the fact that these four sub-units can be combined in so many ways.

The molecules of all kinds of DNA are long thin threads. Although the sequence of A,T,G and C on the threads differs, the molecules all have the same shape.

On the other hand, proteins are constructed from a larger number of different building blocks – the 22 amino acids. There is a lot more variety of shape and size in this set of building blocks than there is in the four DNA sub-units. Therefore their molecules show much greater difference of shape and form than the nucleic acids, and they carry out a wider range of functions, including all the enzyme activities which we discussed in Session Three.

We haven't done any work on the structure of carbohydrates and fats before, so let's have a look at them now.

Carbohydrates include macromolecules composed of sugar building blocks. They are often less complex than the proteins or nucleic acids, because they function as storage compounds and their molecules are composed of only one kind of building block.

Fats are a very varied group of compounds. They are not really macromolecules in the same sense as the others we have been discussing, because few of them are as large, but they do consist of different combinations of a few basic building blocks – in this case, **fatty acids**. These are often combined with other non-fat compounds to give a very varied group of substances with storage and structural functions.

The fact that the macromolecules which form the major part of the food we eat – carbohydrates, fats and proteins – consist of large molecules built up from a basic set of components makes it possible for humans to eat plant and animal foods, break them down into their components, and then rebuild these into the human carbohydrates, fats and proteins from which our bodies are constructed.

Summary

- Our food consists mainly of three groups of substances – carbohydrates, fats and proteins.
- Carbohydrates are made up of sugar building blocks. They are the least complex of the three groups, and often function as storage compounds.
- Proteins are made up of amino acid building blocks. They are a more varied and complex group, having a wide range of functions.
- Some carbohydrates and proteins are very large macromolecules.
- Fats are made up from fatty acid components. Their molecules are not so large, but they often contain additional non-fat elements, giving a very varied and complex group of storage and structural materials.
- Large molecules in food are broken down after being eaten and their components are rebuilt into the human carbohydrates, proteins and fats which make up our bodies.

2: Taking apart and rebuilding

In this introductory unit on biochemistry, it would be unrealistic to attempt an explanation of the nature and characteristics of all three of these groups of food materials. So I intend to concentrate on just one – the carbohydrates. This will illustrate the principle of raw material usage in the cell.

Plants use **starch** as their main store of carbohydrate. This is composed of chains of the sugar **glucose**. Glucose is made by plants from **carbon dioxide** and water, through the process of **photosynthesis**. Its molecules are small, consisting of six carbon atoms, six oxygen and twelve of hydrogen. The structure of a molecule of glucose is therefore described by the formula:

$$C_6 H_{12} O_6$$

where C stands for carbon, H for hydrogen and O for oxygen. The accompanying numbers show how many of each kind of atom are present in one molecule.

Glucose dissolves easily in water – it is **soluble**. If a plant cell has made more glucose than it needs during the hours of sunlight, it can't keep it in one place very easily. Therefore, the plant uses its enzyme machinery to join glucose molecules up to form very large, insoluble molecules of starch, which can be easily stored in the cell.

Each kind of plant makes a slightly different form of starch but all forms consist of very long chains of glucose molecules with occasional branches.

ACTIVITY 21 ALLOW 5 MINUTES

This activity asks you to investigate where carbohydrates such as starch are to be found in your diet.

Today's regulations about food labelling make it easy to find out what our favourite foods contain. I suggest that you look at the labels on your favourite breakfast cereal, or your preferred brand of margarine or butter, to see how much carbohydrate, fat and protein they contain, and to notice anything else that you find interesting.

Commentary

Two examples of labels I found on my breakfast table are shown in *Figure 6*.

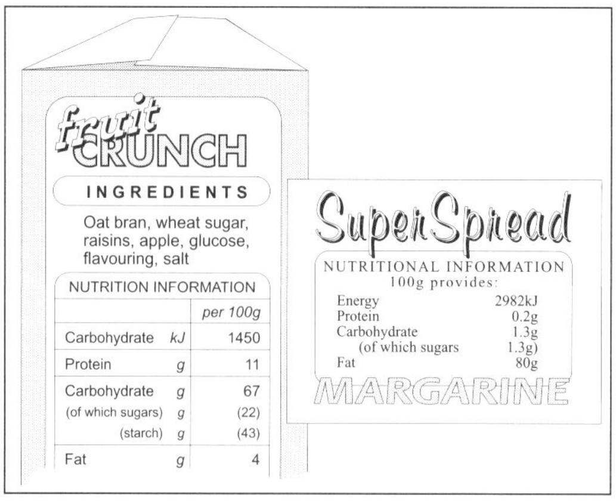

Figure 6 Information from the breakfast table

These labels tell me that my favourite cereal has 67 per cent carbohydrate, all derived from plant sources. The margarine has very little carbohydrate. It is 80 per cent fat. There is no indication of the source of the ingredients.

Both labels tell us how much energy the food contains. You can see that there is a considerable difference between the two foods.

I imagine that the labels you looked at were pretty similar to these. By studying the small print on the food we purchase, it is easy to work out how much of the three main groups of food materials we are eating.

The main carbohydrate in plant foodstuffs is starch, the storage compound described earlier. When these foods are eaten, the starch is broken down during digestion into glucose molecules. They form one of the main fuels for our cellular machinery, as we'll see in Session Five.

But what do we do with any surplus? We have the same problem as the plant. We need to store this soluble material in an insoluble form which is readily accessible when needed.

Animals, including humans, do not produce starch. Our cellular machinery joins glucose molecules together in a slightly different way to produce **glycogen**, a storage material which is better suited for use in human cells. It is more compact than starch, so it can fit more glucose into a smaller volume. Glycogen granules can be seen in many active human tissues, including muscle.

ACTIVITY 22

ALLOW 5 MINUTES

This activity will enable you to identify another way in which cells are adapted to their purpose.

Figure 7 shows a high power magnification of the edge of the muscle fibre first seen in *Figure 1*. Two of the contractile elements – the myofibrils – are shown.

After studying the diagram, you might like to add another adaptive feature of muscle cells to your list in *Table 1* on page 7.

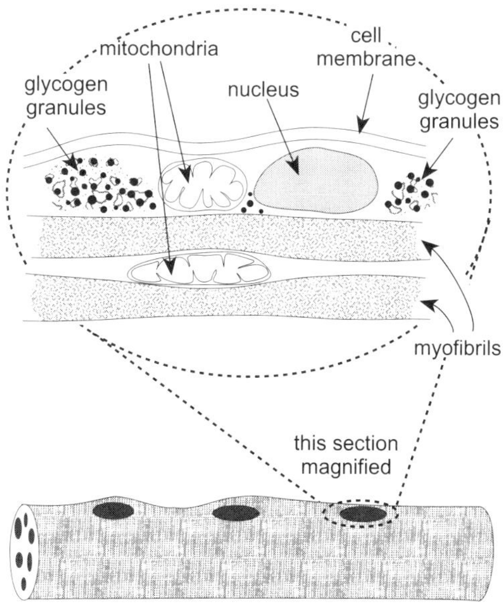

Figure 7 Part of a muscle fibre at very high magnitude

Commentary

This diagram shows glycogen granules arranged next to the contractile fibres. They are there as a store of glucose which is readily available if the rate of muscle activity has to increase. This storage mechanism is another adaptive feature of muscle fibres – they contain a lot more glycogen than many other cells.

This is an example of the use of rebuilt food components. But how are they dismantled in the first place? We can continue using starch as our example to investigate this. In doing so, we will see some more instances of enzymes at work.

ACTIVITY 23

ALLOW **10** MINUTES

This activity shows you how starch is broken down into glucose. *Figure 8* is a schematic diagram which outlines the process of starch digestion.

Follow the process shown in the diagram through, and then explain in your own words what is happening to the starch molecule.

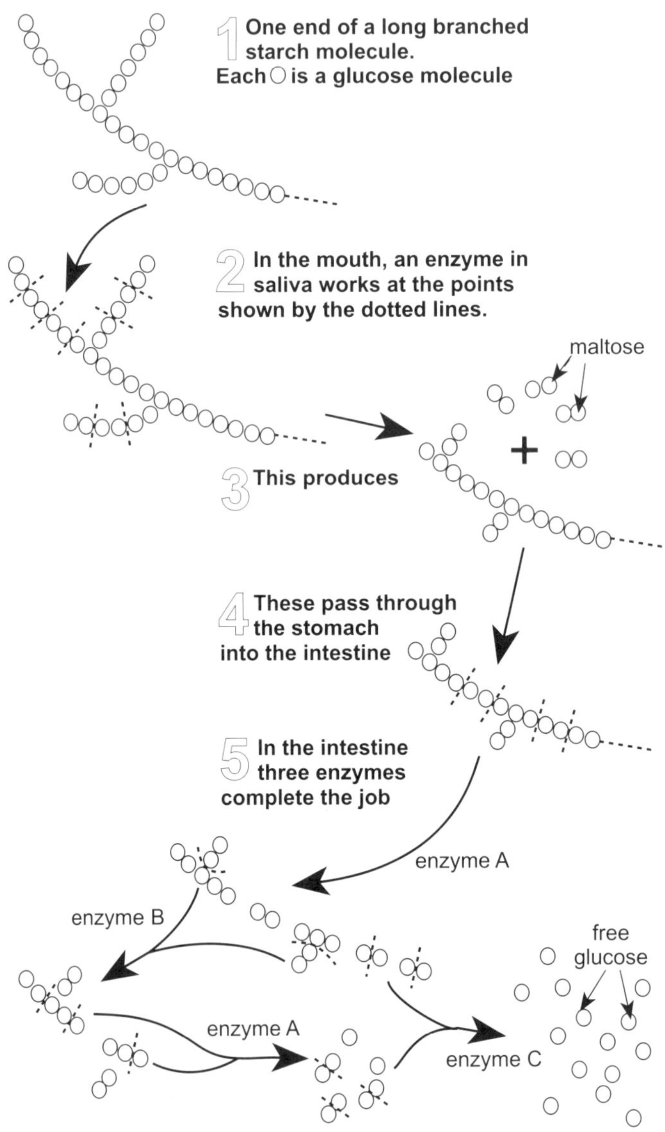

Figure 8 The digestion of starch

Commentary

Your explanation might go something like this.

First, the starch is worked on by an enzyme in the mouth which removes pairs of glucose molecules. These pairs are called **maltose**.

Then the food is swallowed, and nothing else happens to the starch until it has gone through the stomach into the small intestine. Here, other enzymes work on it at different places. One splits the molecule near its branching points; another works on the bits that are left; and a third divides the maltose to leave only glucose.

This series of events shows how the macromolecule is progressively taken apart by enzymes. The various enzymes are specific in their action, each one working on a different part of the macromolecule. For example, the enzyme in saliva works on the bonds which join pairs of glucose molecules together in a straight chain. It does not work on the bonds which join a branch into the main chain. This remains intact until the remains of the starch molecules reach the intestine and encounter enzyme B which works on the branching point. Enzyme C splits the maltose into separate molecules of glucose.

From your earlier work on enzymes, you may have realised that these three enzymes have differently shaped combining sites, which enables each one to work on a different part of the starch molecule.

Although details differ, the same principles apply to the other large food molecules. They are taken apart (digested) by enzymes acting in a co-ordinated way in the digestive tract. The end products of this process are the small molecules of glucose, amino acids and fatty acids.

Summary

- Carbohydrate macromolecules are used for the storage of sugars.
- They are usually simpler than the others as they consist of chains of identical components.
- Glucose ($C_6 H_{12} O_6$) is the building block from which both starch and glycogen are composed.
- Plant carbohydrates are broken down to glucose during digestion by a series of enzymes.
- Glucose is later rebuilt into human carbohydrate such as the storage compound glycogen.
- The same principles apply to the digestion of protein and fat.

All this digestive activity takes place in the digestive system, which is essentially a tube open at each end to the external environment.

In a sense, the contents of the digestive system are still outside the body because they have not passed across the boundaries of any cells yet. They are still on the way to the cellular small business sites. Now I would like to move on to consider how they get to and across the cell boundaries.

3: The route to the factory

Continuing with the example of glucose, let's consider how it gets from the digestive system into the cell where it will be used. This will introduce you to some more kinds of cellular work and some specialist cellular adaptations.

Onto the transport system

From the digestive system, the raw materials such as glucose have to get into the body's transport system – the bloodstream – which will carry them to wherever they are needed. In order to do this, they must cross the barrier of cells that separates the inside of the digestive tract from the blood flowing through the capillaries. *Figure 9* shows a single cell in this barrier.

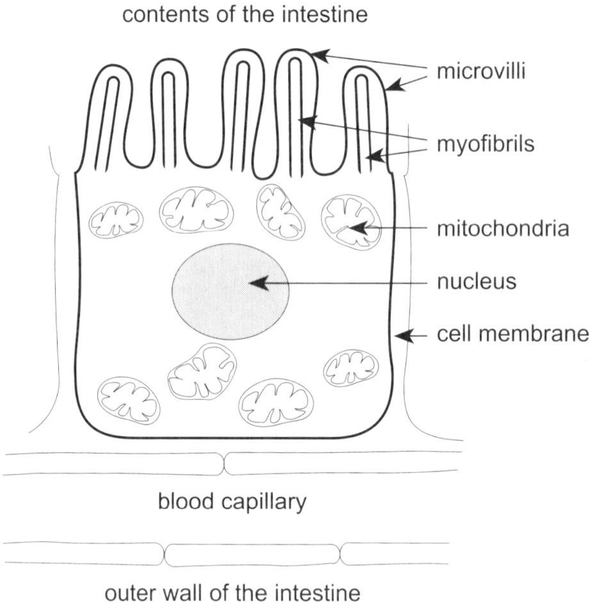

Figure 9 A cell from the lining of the small intestine

This cell has several features which are different from the pancreatic cell we looked at earlier.

ACTIVITY 24 ALLOW 5 MINUTES

The purpose of this activity is to identify ways in which cells in the lining of the small intestine are adapted for their function.

Compare the cell in *Figure 9* with the pancreatic cell shown in *Figure 2*. Identify and note two ways in which the two cell types differ.

-

-

Can you suggest a reason for the presence of the features you have noted in the intestinal cell?

Commentary

Perhaps the most striking difference between the two cell types is the presence of **microvilli** on the intestinal cells. These are projections which cover its whole surface, and their presence produces an increased surface area for absorption of food components such as glucose to take place.

You may also have noticed the presence of myofibrils at the top edge of this cell. These are the same contractile elements which take up most of the space in muscle fibres. They enable the microvilli to move, stirring the contents of the intestine close by and promoting interactions between the digested food and the cell.

The intestinal cell shown in *Figure 9* displays adaptations to allow it to fulfil the function of absorption – taking materials from the digestive system across onto the transport system which will convey them to the cells.

Active transport

The way this occurs demonstrates another kind of cellular work. Glucose, along with other food components, does not drift passively across this cell. There is an active process during which glucose is transported into and across the cell. This is called **active transport** and allows substances to be moved in and out of cells in many different places.

Without an active process, glucose would tend to become evenly spread throughout the cell. The molecules of any substance tend to spread themselves evenly throughout the space available by the process of **diffusion**. This is what happens when, for example, a drop of paint is placed in a glass of water. Even if the water is not stirred the colour will, in time, spread evenly through the water.

Imagine that, for some reason, you were asked to move all the paint to one place in the glass. Experience of the way things work would lead you to say, 'It can't be done'. But in the cell, active transport processes can do something very similar. They can move substances around and in and out of cells against their natural tendency. They can cause a substance to build up in high concentrations in certain cells, even when the natural tendency is for the substance to leak out. This is rather like rolling a stone uphill, and so it requires a supply of energy to power active transport. As you can see from *Figure 9* this cell contains numerous mitochondria at its lower border. These provide the energy to power active transport of digested food units, including glucose, out of the cell and into the blood in the adjacent capillary. We will discuss how they do it later in this unit.

Cells which line cavities and cover surfaces are epithelial cells. The ones in the intestine perform several kinds of work, including active transport. They work in a rather hostile environment and they are relatively short-lived. An absorptive cell like the one in *Figure 9* will only live for about five days. It is then shed and replaced by a new cell which has grown from below. The whole lining of the digestive tract is constantly being replaced by cell division.

ACTIVITY 25

ALLOW 5 MINUTES

The purpose of this activity is to relate what you have learned about the process of food absorption to the effects of chemotherapy.

In *Activity 8*, I asked you to think about protein manufacture in several people, including a young man undergoing chemotherapy. As you may know, chemotherapy can produce digestive disturbances and soreness in the mouth, as well as reduction in the ability to use food.

Using all the information you have, can you suggest a biochemical explanation for these effects?

Commentary

Anti-cancer drugs often interfere with cell division. This stops the growth of cancer cells, but may also stop the division of healthy cells. If this happens, then it will also affect the replacement of cells in the digestive system, right through from the mouth. It also reduces the number of absorptive cells, so the raw materials from food can't be actively transported into the bloodstream.

We have taken just a brief look at the first barrier across which the raw materials must pass on their way to the cells. Once in the bloodstream, the raw materials are rapidly transported to all parts of the body.

In the tissues, the plasma leaves the terminal capillaries and forms the intercellular fluid which surrounds every cell. The small molecules of glucose, amino acids and fatty acids are transported to the cell surface in this fluid. They have arrived at the factory gates.

Summary

- Cells covering surfaces and lining tubes are termed epithelial cells.
- In the small intestine, the epithelial cells are adapted to absorb the components from digested food into the bloodstream.
- This is not just a passive process of diffusion. It involves active transport.
- Active transport processes require energy, which is provided by the mitochondria. These are plentiful where active transport is taking place.

4: The factory gates

As you can see from all the diagrams of cell structure in this unit, the boundary of a cell is the **cell membrane**. This is much more than a simple boundary, however. It is covered with structures which control the passage of different materials in and out of the cell and which are involved in active transport. Many of these are themselves controlled by external agents, such as insulin. For instance, when there is plenty of glucose in the blood, the production of insulin increases. The hormone travels round in the bloodstream, opening the 'gates' in cell membranes and allowing glucose into the cells.

How this actually happens is not yet fully understood. However, it is known that there are places on the surface of the cell membrane where hormones, including insulin, become attached. These are termed **receptor sites**, and they are specific for different hormones, in the same way that combining sites on enzymes are specific for their substrates.

When an insulin molecule is attached to a receptor site it causes a change in the cell membrane which allows glucose to enter the cell. It really does 'open the gate'. When insulin is in short supply for any reason, the gates remain closed. If this situation continues for any length of time, the body is unable to use glucose properly and needs to get rid of the excess from the blood. This gives rise to the symptoms of diabetes.

ACTIVITY 26

ALLOW 5 MINUTES

Using another case study, this activity shows you how the work you have done so far can help you to understand the biochemistry of diabetes.

I would like to introduce you to Jason at this stage, and provide more information about him as the unit progresses. In this way, I hope to build up your understanding of the biochemistry of his particular illness. Please read the first instalment and consider the question which follows it.

Jason is a university student in his second year of an engineering course. He is 19 and has recently been unwell. His weight has dropped by over a stone in the last six months, he feels continuously thirsty and is passing large amounts of urine. He reluctantly visits the student health centre where, after taking a blood test, the doctor suspects that he has early-onset diabetes mellitus and arranges for him to be seen quickly by a physician.

Assuming that this provisional diagnosis is correct, which of Jason's symptoms can you explain in terms of a shortage of insulin in his body?

Commentary

On the basis of the work we have done so far, you would probably be able to explain one of these symptoms – the weight loss. Since glucose will be unable to get into his cells, he will be unable to use his food properly and will therefore lose weight.

The other symptoms, thirst and large volumes of urine, are due to the build-up of glucose in the blood, as we will see later.

First, let's see how Jason's condition differs from that of a person without diabetes.

When insulin is produced normally, it opens the factory gates on demand, as it were. That is to say, when the blood sugar level is high, insulin production is stimulated and more of the gates open, so that glucose is taken up, restoring the normal blood sugar level. This applies to most types of tissue, but not to the brain.

allow as long as you can spare

ACTIVITY 27

I'd like you to try a simple experiment this time, to see if you can detect any of the effects of your own changing blood sugar level.

Choose a day when you're not at work and can concentrate on this activity over the course of several hours. At twenty minute intervals, notice how alert you feel. At the same time, note down any other observations, such as whether you are yawning, have a mild headache, or feel irritable. Record your responses for as long as you can manage in *Table 2*.

If you feel very alert or sleepy, put four crosses in the appropriate column. Use more or less crosses to indicate intermediate stages. Also note times at which you have anything to eat or drink.

TIME	F E E L I N G S					
	Sleepy	Alert	Yawning	Headache	Irritable	Other

Table 2 Your observations

Commentary

Here's an example of the experiment I did on New Year's Day, using myself as the subject. How does it compare with yours?

TIME	FEELINGS					
	Sleepy	Alert	Yawning	Headache	Irritable	Other
13.20	+ +			+		
13.40	+			+		
14.00	+			+		
Lunch						
14.20		+ +				
14.40	+ + +		+ +	+		
15.00		+				
15.20		+ + +				
16.00		+ + + +				
16.20		+ + +				
16.30	+		+ +			
16.40	+ + +		+ + +	+		
Two cups of tea						
17.00		+				
17.20	+					
17.50		+ +				
18.00	+			+		
18.20		+				

Table 3 My results

In *Figure 10*, I have also drawn the sleepy/alert levels as a graph, to see how they relate to mealtimes. You might like to do the same.

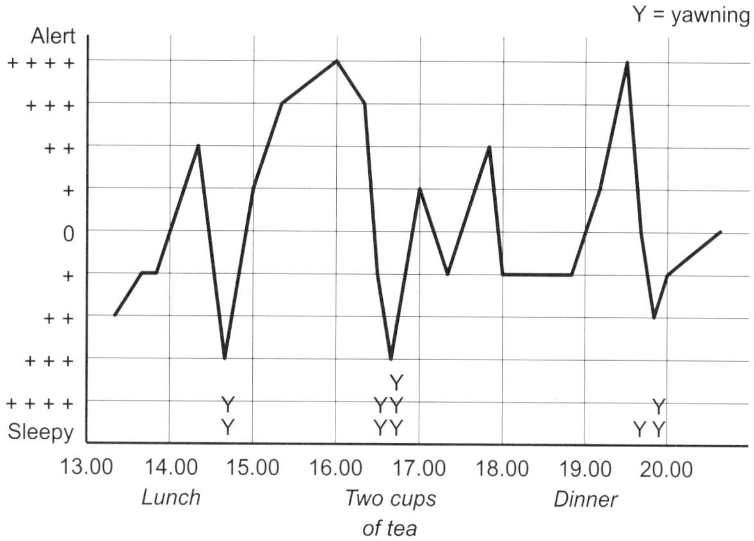

Figure 10 Alertness and sleepiness during one day

From the results, it looks as if the degree to which I felt alert on that day could be related to food, although there may be other factors; one day's figures are not conclusive. It is probable that the way I felt was related to blood sugar levels. It is well known that shortly after eating foods containing sugar, the blood sugar level rises and insulin production is stepped up. This mops up a lot of the sugar and may overshoot, producing a transient fall in the blood sugar level and leading to drowsiness. After a short time, the balance between insulin production and blood sugar is restored and the level stabilises.

You may find it interesting to chart your own response over a number of days in order to see if you can detect a pattern. This can be helpful in all kinds of ways when planning activities, and can also help us as health carers to work with patients and clients in planning for their particular circumstances.

It is particularly necessary to understand and to be able to measure how blood glucose fluctuates in people with diabetes, as the next activity demonstrates.

ACTIVITY 28

ALLOW 10 MINUTES

Resource 5 contains a graph showing how blood glucose varies during the day in a person with diabetes. You can see that the general shape is very similar to *Figure 10*, but the fluctuations go well above and below the band marked 'normal'. *Resource 5* also provides another example of the use of enzymes in health care, this time for the purpose of testing blood glucose levels.

Now read through the article and record your answers to the following questions.

1 What is the normal range of values for blood glucose?

2 How are enzymes used in the testing process described in this paper?

3 What unfavourable conditions can affect the activity of these enzymes?

4 What does the paper tell us about the feelings of people with diabetes and the need for regular and accurate testing?

Commentary

1 The normal range given here is 3.5 to 6.5 millimoles per litre.

2 Test strips for blood glucose are impregnated with enzymes which produce a colour change when they react with glucose.

3 If the strips have been stored for too long or have been kept warm, the enzymes may lose activity in the same way as the enzymes streptokinase and streptodornase, which we investigated in *Activity 17*.

4 The paper mentions several responses made by people with diabetes to the need for regular testing – for example, embarrassment, resentment, lack of confidence and simply feeling fed up with the whole business.

This paper clearly shows some of the ways in which health carers can support people with diabetes and their families, and how an understanding of the biochemical basis of this condition can be used to explain the need for accurate monitoring of blood glucose.

Although I have described how glucose gets into the cells, I haven't yet indicated why the glucose supply is so important and, indeed, why it should make such a difference to our well-being. As we'll see in Session Five, glucose is the main source of energy for many of our tissues.

The brain in particular uses only glucose as its energy source under normal circumstances. Muscles and the liver store glucose as glycogen for emergency use. Most of the work done in the body depends on a steady level of circulating blood glucose. So the maintenance and control of this level is of great importance to our well-being.

ACTIVITY 29

ALLOW 5 MINUTES

To conclude Session Four, this activity should help to bring together all the work you have done in this session.

As you probably know, athletes who take part in endurance events prepare by eating carbohydrate-rich food, like pasta.

I would like you to think for a moment about such an athlete who, the night before a big race, has eaten a large plate of pasta. Imagine a molecule of glucose locked up in the starch from which the pasta was made. Note down what is going to happen to that glucose molecule during the next 24 hours, until it is used in the athlete's muscles during the race.

You might like to draw a simple diagram to summarise what you know about the route it will take. Don't worry if your understanding is incomplete – just include the stages that we have discussed in this unit.

Commentary

Your version may look different from mine, but I hope you have managed to incorporate all the stages I have indicated in *Figure 11*. This shows what will happen to a glucose molecule as it is first liberated from the plant carbohydrate storage, starch, and then built up into human carbohydrate storage, glycogen, until it is needed by the exercising muscle. Then it is liberated from the glycogen for use as fuel for muscular movement.

All the carbohydrate food we eat is ultimately converted to glucose and used in this way – either directly as fuel or placed into storage. There is, however, only a limited space for storing glycogen. If you look back at the diagrams of muscle fibres, you will see that they are packed full of contractile myofibrils and the glycogen granules are squashed around the edges. The liver can hold more glycogen than the muscles, but the storage space is limited there too. If the diet contains more carbohydrate than is needed for energy and all the glycogen storage is full, then the remaining carbohydrate will be converted to fat, as we will see in Session Seven.

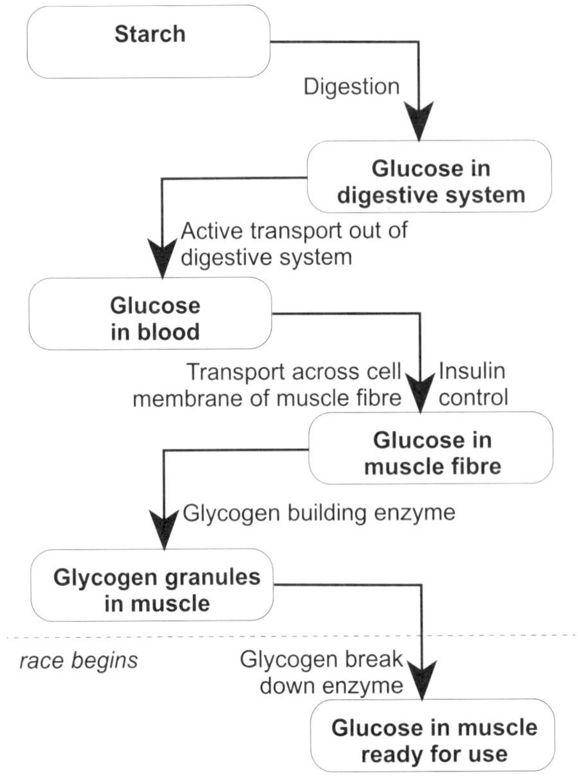

Figure 11 The path of glucose to exercising muscle

We have concentrated on carbohydrates in this session because they are the least varied of the macromolecules in our food, and therefore the way they get to the cell is easier to describe. Whilst presenting a more complex overall picture, fats and proteins are subject to a similar overall sequence.

Fats from all sources are taken apart in the digestive system to produce fatty acids. These are transported around the body in the blood and enter cells through their own 'gates', either for use as structural and energy sources or for storage. Proteins yield amino acids which follow similar pathways into cells. The amino acids are used as raw materials for building human proteins of all kinds, including enzymes. Our bodies do not have a method for storing excess amino acids so these have to be removed. So although the details of all this activity are enormously complex, underlying them is a simple and elegant pattern of dismantling, transporting, storing and using raw materials in the cell.

Summary

- Glucose from digestion travels round the body in the blood plasma.
- The blood glucose level is maintained at a constant level in healthy people.
- Cells have insulin receptors on their surfaces.
- Insulin becomes attached to these receptors and opens the glucose 'gates' when the blood glucose level is high.
- Low levels of insulin prevent glucose from entering the cells and being used. The symptoms of diabetes ensue.
- Muscle cells store glycogen for a readily available supply of glucose to act as fuel for their work.
- All food macromolecules undergo a similar series of steps in which they are dismantled and the components rebuilt, used or stored.

Before you move on to Session Five, check that you have achieved the objectives given at the beginning of this session and, if not, review the appropriate sections.

SESSION FIVE

The power supply

Introduction

A crucial requirement for any small business is a cheap, reliable power supply to enable it to carry out many diverse activities. These can range from the use of information technology to the operation of heavy industrial plant, and include all the maintenance operations like heating, lighting and repairing the premises.

Similarly, the cell carries out many kinds of work and needs an efficient power supply in order to do so. In this session, I will first discuss some of the kinds of work carried out in all cells and some of the specialist work done by a few. Most of the session is concerned with the power supply which makes this work possible, where it is located and how it works. Finally, we will consider the efficiency of this system and how the cell deals with the by-products arising from its power supply.

Session objectives

When you have completed this session, you should be able to:

- describe where the energy to drive cellular activity comes from
- explain what mitochondria are and how they function
- explain where mitochondria are thought to have originated.

1: The need for energy

Energy is a difficult thing to describe. We can't observe it directly; we can only observe its effects. Energy enables work to be performed.

In everyday life, we are involved in or affected by many different kinds of work, all of which require an input of energy. If you boil a kettle, drive a car or switch on a light, work is being performed. As you read this text, your body is engaged in many different kinds of work, even if you're sitting quite still. I would like to begin this session by considering some of the forms of work which cells undertake.

Varieties of work

In order to introduce cellular work, perhaps we could identify some everyday kinds of work and energy.

ACTIVITY 30 ALLOW 5 MINUTES

You can either think through this activity or actually carry it out.

Suppose that you have decided to take a break from work at this moment. You get up, go into the kitchen and make a cup of coffee. During this process, you will carry out several different kinds of activity or work and use household equipment to do other kinds. Make a short list of those that you can identify and the source of energy which makes that work possible.

Type of work	Source of energy

Commentary

You'll have made the coffee in your own particular way but, without doubt, some of the following kinds of work will have taken place.

- Mechanical work: getting up from the chair and moving around the house requires an input of muscular energy.
- Transport work: carrying objects such as cups and pans around also requires muscular energy. It is another form of mechanical work.

- Thermal work: boiling a kettle or heating a pan both require an input of heat energy, supplied either by electricity or by burning a fuel such as gas.
- Chemical work: as you stir or percolate the coffee, it undergoes a change. The aroma becomes stronger and the solid material dissolves in the liquid. This also needs thermal energy.
- Electrical work: if you switch on the light, electrical energy is converted in the light bulb into the radiant energy of visible light.

All these kinds of work need an input of energy from some source. In each case, the source you have identified is not the original source of the energy. All forms of energy which we use to do everyday work represents a **conversion** of the original source into something which we can use.

For example, suppose that the electricity you use is produced in a hydroelectric power station. Where is the original source of the energy? One answer is in the mass of water confined behind a dam. As it is released, it causes turbines to rotate and generate electrical energy. We could not make direct use of the water in the dam to boil a kettle, but we can use it after its energy has been converted to another form.

Energy conversion

If you think about the power station example again, you may well conclude that the answer offered is only a partial one. It still leaves the question of how the water came to be behind the dam in the first place – and therefore how it came to have the energy which propels it through the turbines.

The more complete answer is that its energy came from the sun, which raised water vapour from the earth's surface into the clouds, from where it fell as rain in the hills. Its presence in the dam is the result of a series of **energy conversions**, starting in the sun.

Almost all the energy we use arises from the same source – **solar energy**, which enables plants to grow.

This is converted into the **chemical energy** which holds atoms together in the macromolecules made by plants. Over time, the plants die and decay to form fossil fuels – oil, coal and gas. The chemical energy from the plant material remains in the chemical structure of these fuels. When coal is used in a power station, its chemical energy is converted into electrical energy which, in its turn, can then be converted into other types such as mechanical energy or light.

The human diet contains material from both plant and animal sources. Animals rely on plants, either directly or indirectly, for their food. So the solar energy which forms the chemical bonds in plant materials is the source of all the chemical energy in our food.

These relationships are shown in *Figure 12*.

You may notice that all the kinds of energy shown in this diagram can be traced back to the sun – to solar energy. At various stages in the pathways, one kind of energy is changed to another. Such a step is termed energy conversion. These steps are of the utmost importance because they allow a natural source of energy – the sun – to be used to sustain life.

Human inventiveness has produced many **energy conversion systems**. We can convert the chemical energy of petrol into mechanical energy via a car engine and

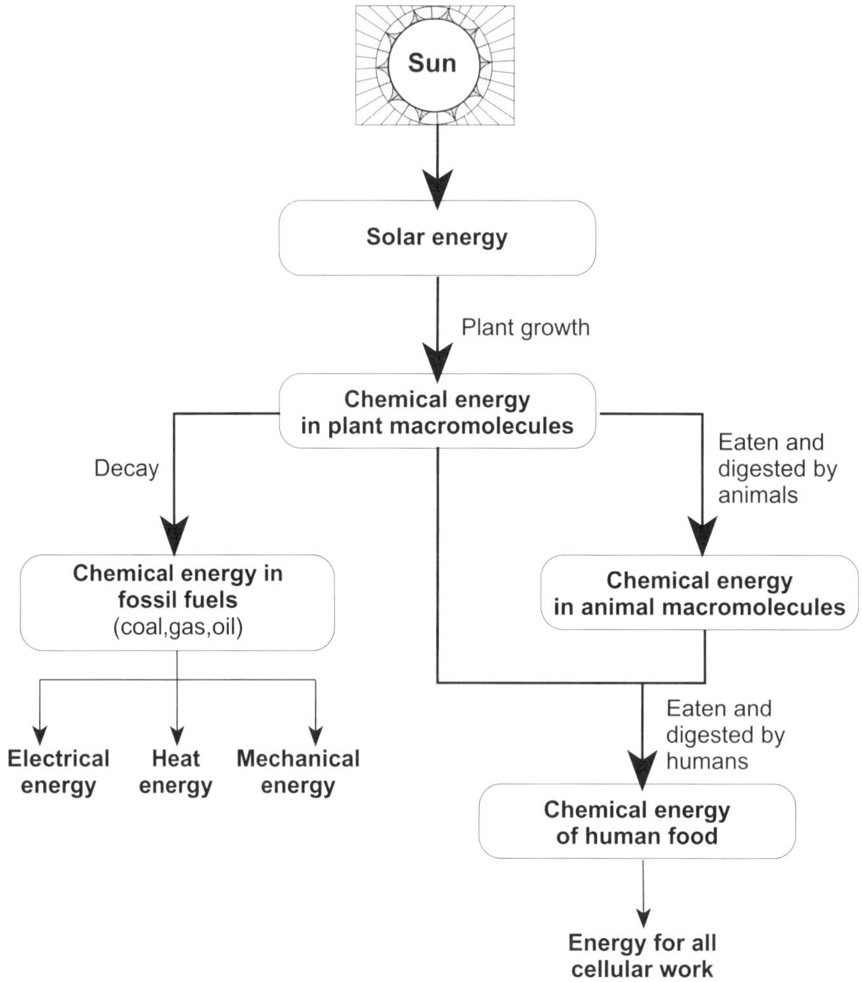

Figure 12 The relationship between sources of energy

electrical energy into light by means of an electric light bulb. Many of the pieces of equipment that we use at home or at work are energy converters. They change one form of energy into another which is useful for a particular purpose. Without an energy converter the chemical energy in petrol couldn't help us move from place to place, nor would the electrical energy passing through a wire help us to see in the dark.

When an energy converter operates, it is never 100 per cent efficient. Only a proportion of the energy that goes in as one form emerges as the useful output form. The rest is wasted, usually as heat – which can be thought of as 'useless' energy, unless it is being deliberately produced by the system. In a car engine, for example, the heat must be removed by the cooling system – but some of it is useful as a means of heating the interior of the car on cold days.

Living cells also have to convert energy. Their primary source is the chemical energy in food materials which, as shown in *Figure 12*, originates in solar energy. Living cells have to convert this chemical energy into a useful form and to deal with the waste heat produced.

The **mitochondria** are the structures in which this energy conversion takes place, and we will describe them in detail later. But before that, I would like to review the kinds of work for which the cell needs this useful energy, since this will increase your understanding of the problems which the living energy converters must solve.

Work in the cell

In *Activity 30*, you identified several kinds of work that are familiar in everyday life. Now let's see if there are parallels inside the living cell.

ACTIVITY 31 ALLOW 5 MINUTES

Looking back at *Activity 30*, list the kinds of work which were identified there. Next, using the knowledge you have so far acquired in this unit, decide if cells carry out a similar type of work. You might like to record your answers in *Table 4*.

In some cases, you may already be able to state how cells carry out a particular type of work. If so, include this in the table. If not, put a question mark in pencil next to the item.

Type of work	In what way, if any, do cells do this type of work?
Mechanical	
Transport	
Thermal	
Chemical	
Electrical	

Table 4 Work done in the cell

Commentary

At this stage in the unit, you should have realised that muscle cells move by contracting, so they do mechanical work. Transport is carried out by red blood cells, which use haemoglobin to carry oxygen around the body.

All cells do chemical work, as they use enzymes to bring about chemical change. Thermal and electrical work may have been a bit more problematic, but you should be able to complete these items later on.

ACTIVITY 32 ALLOW 10 MINUTES

Are there any other forms of cellular work which we could add? Look back through the unit to see if you can identify any other kinds of cellular work that have been mentioned, and add them to *Table 4*.

Commentary

In Session Two we concentrated on what the ribosomes do within the cell. Their work is to manufacture proteins, so you could add building up proteins to your list in *Table 4*. We also mentioned building messenger RNA, and in Session Four we talked about building carbohydrates and fats. The general term used in biochemistry for building up large molecules from their component parts is synthesis. This is a type of chemical work.

We have described one way that substances get across cell boundaries, in Session Four, as active transport. This too requires energy and is a form of work, so it should be another addition to your table, in the transport section.

Summary

- Energy enables work to be performed.
- There are many different forms of work and energy.
- Almost all the energy we use in our daily lives has its origin in solar energy.
- Energy may be changed from one form to another through an energy conversion system.
- The object of energy conversion is to produce a form of energy which can be used for a specific purpose.
- Living cells must carry out energy conversions. Their energy comes, ultimately, from the sun.
- Cells carry out energy conversions to produce forms of energy which can be used to do cellular work.

We are starting to build up a picture of the living cell as a small business, carrying out many kinds of work. It needs energy supplies for all of them. The remarkable thing is that living cells use a single energy source for all these diverse activities – just one energy source powers all these kinds of work. Let's look at this next.

The cell's energy source

Under the conditions existing in the human body – that is, under normal physiological conditions – our cells can carry out many different kinds of work. You've already listed some of these in *Table 4*.

All these sorts of work need energy. The single form of energy used by our cells is a substance called **adenosine triphosphate (ATP)** which is uniquely able to provide the energy for a wide range of cellular activities and to be regenerated after use within the cell itself. It is a truly wonderful and effective source of energy.

ATP allows things to happen which would normally be physiologically impossible. For example, to start to make a protein artificially, we would need to join together two amino acid molecules in the laboratory. This requires heat and the presence of other chemicals, and takes a long time. But enzymes on the ribosomes can join amino acids together very quickly at body temperature, as long as they have some ATP. In the cell, the ATP supplies the energy, whereas in the laboratory, the biochemist needs to heat the reaction mixture.

ATP is a small molecule which consists of four parts. We don't need to consider the details of its structure; we can simply visualise it as having –

- a main section: the **adenosine** (A) portion

with

- three smaller sections: the **phosphate** (P) portions

– linked to the main section. Take a look at *Figure 13*.

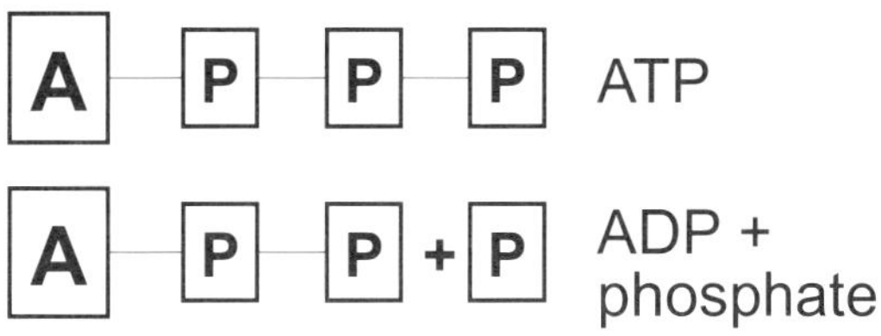

Figure 13 A model representing ATP

When ATP is used to provide energy for cellular work – to join amino acids together during protein synthesis, for example – it loses the last phosphate in the chain of three and becomes **adenosine diphosphate**, or **ADP** for short, as in the second part of *Figure 13*. ADP cannot be used as an energy source, but it can be built up again into ATP when the conditions are right.

When the third phosphate of ATP is released, a large 'bundle' of energy is set free. This bundle can be picked up by the working parts of the cell, and used in all the kinds of work we have considered so far. The ATP energy is channelled into cellular work.

ACTIVITY 33

ALLOW 2-3 MINUTES

The purpose of this activity is to help you construct a mental picture of how ATP functions. It isn't easy to find a familiar analogy for this process, but I hope the following will help.

Think of an occasion when you have been involved in a piece of work with another person who needed to be encouraged or motivated. It may have been a child, a colleague, or a patient or client who needed a lot of input from you to carry out their task. Reflect for a moment on how it felt and then, in a couple of sentences, jot down your observations of the event.

Commentary

Several people who tried this activity said that they felt 'drained' after putting a lot of effort into working with a person whose motivation for a particular task which had to be done was low. They felt that they were supplying all the energy needed both for themselves and the other person. Afterwards, they needed to be 'recharged' before they could give out again. Sometimes in health care, people give out so much in terms of physical and emotional energy that they suffer 'burn out', and need a long period of recharging.

The idea of 'giving out' to make things happen is one which we can apply to the way that ATP functions. This unique molecule has a structure which allows it to give out energy in just the right way to bring about work in the cell. In so doing, it becomes 'burnt out' (converted to ADP) and needs to be recharged before it can give out again.

I would like to try one more analogy to make the point clear. Imagine a see-saw with a person standing on one end. If someone else climbs to the top of a high wall nearby and jumps onto the other end of the see-saw, what will happen?

The first person will fly up into the air. Circus acrobats make use of this effect in some of their acts. Their jumps are controlled, just sufficiently to send the person on the other end of the see-saw flying off to the right height and distance. When the acrobat jumps, just the right amount of energy is released to produce the desired effect. ATP molecules do exactly this, so that cellular work is done in a controlled way.

ATP is used in all kinds of work, as I have already emphasised. But it is also produced from all the raw materials which enter the cell – the chemical energy from food is converted to ATP. So the cell is a lot more efficient than the small business, because it can satisfy all its energy needs by the production and use of this single source, instead of dealing with many different ones such as electricity, gas, petrol and so on, which involve numerous energy conversions.

Summary

- Living cells perform many different kinds of work.
- ATP is the energy source for all of them. It is produced from all kinds of food.
- After providing energy for cellular work, ATP is converted to 'useless' ADP and must be recharged before it can be used again.

Now that you have some information about the nature of the cell's energy source and the way that it is used, we need to consider how it is formed in more detail. To do this, we must go back to just inside the 'factory gates' and see what happens to the raw materials which are brought in to the cell, as it is their chemical energy which is destined to become ATP energy.

2: The fuel supply

Using glucose

In the previous session, we looked at how food macromolecules are broken down into their basic components. These components are then available for use by the cell, and they may:

- be synthesised into cellular products
- stored

or

- used as fuel for the mitochondrial power stations.

Here, I would like to consider the third of these options – what happens when the food components are used as fuel.

Glucose is the main supply of fuel for the cell. As you have already seen, it is produced when carbohydrate macromolecules are broken down during digestion.

When carbohydrates are digested to glucose, some of the energy holding their components together is set free as heat, which is used to maintain body temperature. It is used to perform thermal work. However, most of the original energy which formed the macromolecule remains locked up in the bonds holding together the atoms in the glucose molecules. When glucose enters the cell, it undergoes a process of rearrangement which enables that energy to be passed on to form ATP. Let's look in more detail at what happens to glucose during this process.

Splitting glucose

First of all, a series of enzymes working in sequence act on the glucose.

These enzymes change each glucose molecule into two molecules of **pyruvic acid**. Each glucose molecule has six carbon atoms and each pyruvic acid molecule has three.

We have summed this up in a single sentence, but it is actually a very complex series involving ten separate steps, each one carried out by a specialised enzyme. The sequence is called **glycolysis**, which literally means 'sugar splitting' because the six-carbon sugar molecule is split into two three-carbon molecules. The energy released in this splitting process is used to make two molecules of ATP for each molecule of glucose. In addition, each glucose molecule loses four atoms of hydrogen.

Equation 1

$C_6H_{12}O_6 + 2ADP + 2P$	$\rightarrow$	$2 \times C_3H_4O_3 + 2ATP + 4H$
(glucose)	(enzymes of glycolysis)	(pyruvic acid)

You will see from this equation that oxygen doesn't play any part at this stage. Glycolysis does not need any oxygen. It allows the cell to produce a small quantity of ATP in the absence of oxygen. We'll have a practical demonstration of this in the next activity, but first let's summarise some of the key facts about glycolysis, and explain what this equation means.

Summary

- In order to make use of the energy in the glucose, its molecules must be rearranged so that the energy can be handed on to ATP.
- In the cell, the first stage in the use of glucose is a series of ten steps, each brought about by a specialised enzyme.
- The overall process is called glycolysis. It does not require the presence of oxygen.
- Glycolysis results in each glucose molecule being split into two three-carbon molecules (pyruvic acid) and releasing four atoms of hydrogen.
- Glycolysis results in the production of a small amount of ATP. (Two molecules of ATP are gained for each molecule of glucose used.)

Using equations

Equations such as *Equation 1* are used to convey information about chemical and biochemical change. The essential feature of an equation is that it balances. In other words, there must be the same number of atoms on each side of the arrow.

This equation tells us that each molecule of glucose splits into two molecules of pyruvic acid. A molecule of the latter has three atoms of carbon, four of hydrogen and three of oxygen. So two molecules of pyruvic acid account for six carbon, eight hydrogen and six oxygen atoms. Four hydrogen atoms are left over from the glucose molecule, which we write as 4H – indicating that these atoms have been liberated from the glucose in some way.

The equation also tells us that, during the splitting process, two molecules of ADP have been recharged to form ATP. Some of the energy holding the sugar molecule together has been used to form ATP from ADP and phosphate.

Please remember that *Equation 1* is a summary of ten separate steps in which the molecules of glucose are rearranged in order to allow this to happen. All ten steps are necessary if this part of the sugar's energy content is to be used efficiently and not wasted.

Working without oxygen

Glycolysis doesn't need any oxygen. It produces a small amount of ATP for each glucose molecule, and it produces free hydrogen atoms.

In the absence of oxygen, cells can produce a supply of ATP through glycolysis, which enables them to carry on working, as the next activity demonstrates.

ACTIVITY 34 — ALLOW 10 MINUTES

This activity shows you what happens when cells work without oxygen. This really is a physical activity, so please only try it if you are fully fit.

First, raise one arm straight up in the air above your head, and let the other hand rest by your side. Clench both fists tightly, and then open out your fingers wide. Repeat this at the rate of once or twice a second. Try to keep clenching both fists at the same rate. Keep going for about five minutes, and record what you observe.

Stop and rest for a minute. Then try again, with the opposite arm raised this time. Again, record your observations.

Commentary

You probably noticed that the arm raised in the air became tired first, and that it was difficult to keep up the same rate of activity. Some people find that their dominant hand – the one they write with – is stronger than the other one in this exercise but that, even so, it is much harder to keep going with this hand when it is raised than when it is dependent.

Why should this be so? We could say that the muscles are getting tired, but why should they get tired more quickly when the arm is raised?

The answer is that, by clenching and unclenching the fist rapidly, we are making muscles work hard and using up the available ATP which is supplying the energy for that work. As long as the blood is bringing oxygen to these muscles, they can work at this rapid rate. But if the blood cannot bring so much oxygen, once the existing supply is used up we are making the muscles work in the absence of oxygen – in other words, **anaerobically**. Under these circumstances, glycolysis can take place producing a small energy output.

This allows the muscles to keep going if the exercise is not too violent, but it incurs a penalty. Do you remember those four hydrogen atoms in the glycolysis equation? During glycolysis, more and more hydrogen atoms accumulate. They clog up the system and, unless they are removed, it will seize up completely. Before this happens, as a last resort, the free hydrogen atoms can be transferred to pyruvic acid itself. Each molecule can accept two hydrogen atoms to produce lactic acid, as shown in *Equation 2*.

Equation 2

$C_3H_40_3+2H$	$\rightarrow$	$C_3H_6O_3$
(pyruvic acid)	(lactate dehydrogenase)	(lactic acid)

This allows the muscle to continue to work for longer still but, in time, the lactic acid itself will accumulate. It is a by-product of glycolysis, and alters the environment in which the enzymes of glycolysis work. When its level in muscle fibres begins to rise, these enzymes slow down until finally glycolysis stops, no more ATP can be produced and the muscle cannot contract any more. If you continued *Activity 34* for a longer period, you would find that it eventually became totally impossible to clench your raised hand because the muscle could no longer contract.

Lactic acid can easily pass out of muscle cells and into the bloodstream, so it is removed from exercising muscle where the blood supply is good. However, it begins to build up in the plasma, lowering its pH and producing **acidosis**. This can arise from several causes other than exercise. For example, if the blood supply to a tissue is poor, lactic acid can accumulate. Or in diabetes, excessive use of fat as an energy source produces ketones, which are also acidic and easily diffuse into the bloodstream.

If the plasma becomes more acidic in this way, it liberates carbon dioxide (CO_2) from one of the minerals in the plasma – bicarbonate.The result is that the blood then carries an extra load of carbon dioxide gas. This is detected by the nervous system and the rate of breathing is increased to enable the carbon dioxide to be blown off from the lungs.

When a person exercises vigorously, their breathing rate increases for these reasons. You may have noticed that when you have been exercising you continue to breathe hard after you have stopped. This is because the lactic acid is still present in your blood until it has been disposed of in the liver and you need to continue to get rid of carbon dioxide.

It is important for nurses to understand what is happening in the muscles and bloodstream when caring for a person with respiratory problems. When the oxygen supply in the blood is limited by conditions such as chronic bronchitis, emphysema and cystic fibrosis – all conditions which reduce the amount of oxygen the blood can carry – then there will be a shortage of oxygen in the muscles, and sustained muscular effort will be impossible.

Summary

- Glycolysis can occur anaerobically – that is, in the absence of oxygen.
- Cells can carry on working without oxygen, using the small amount of ATP produced in glycolysis.
- However, if the cells have to work hard, the hydrogen atoms produced in glycolysis build up.
- They can be mopped up by pyruvic acid, but this is only a temporary remedy, as this will produce lactic acid which slows down enzyme activity if it accumulates.
- Lactic acid passes out into the bloodstream. This causes acidosis and an increase in the rate of breathing.
- People with respiratory problems may have difficulty getting enough oxygen to their cells, so may have to rely on anaerobic ATP production. Sustained activity is very difficult under these conditions.

Into the 'power station'

We have described what happens to glucose in the absence of oxygen. Most of the time there is plenty of oxygen, delivered regularly to our tissues in the blood. Under these circumstances, the pyruvic acid and the hydrogen produced in glycolysis enter the mitochondria, where most of the chemical energy from the original glucose molecules is used to make a lot more ATP. It is important to emphasise that these two products of glycolysis, pyruvic acid and hydrogen, still retain most of that original energy.

The mitochondria are often called the 'power stations' of the cell. They are its main energy converters, because they take in pyruvic acid and hydrogen, which carry chemical energy from glucose, and produce ATP, which is the energy source that the cell can use.

When oxygen is plentiful in the cell, the 'gates' of the power stations open to admit the products of glycolysis.

Next, we will take a detailed look at the mitochondria to see what happens there.

3: Mitochondria – cellular 'power stations'

Where mitochondria occur

Figure 1 on page 8 shows two specialised types of cell. Take another look at it now – you will see that each cell contains round or oval structures labelled mitochondria. These are sub-cellular structures concerned with the conversion of chemical energy derived from food into a form of energy which cells can use in all the kinds of work that they do.

Mitochondria occur in all living cells, and the more active the cell – the more work it has to do – the more mitochondria it will contain. Very active cells contain thousands of mitochondria.

They tend to accumulate in the parts of the cell where most activity takes place.

For example, sperm are very active structures. They are sometimes referred to as 'naked nuclei', containing very little apart from the genetic material.

Figure 14 shows that, in fact, the sperm contains something else as well, to aid it as it moves towards the egg. It contains a stack of mitochondria arranged at the base of its tail, to provide the energy it needs in order to sustain movement as it travels towards the neck of the womb.

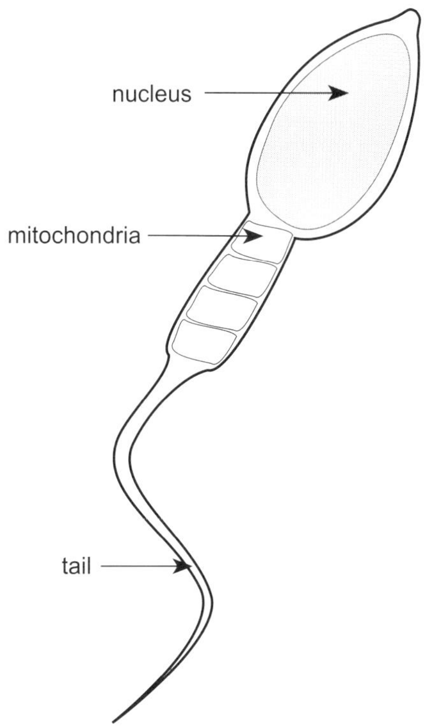

Figure 14 Diagram of a single human sperm

Mitochondria are also plentiful in all the specialised cells which we are using as examples in this unit. These are all very active, but in different ways. Muscle requires energy for movement and heat production. Pancreatic cells need energy for the synthesis of insulin, and intestinal epithelial cells need energy for active transport.

Young mammals, including human babies, have a specialised tissue called 'brown fat'. This is a form of fatty tissue which has a very high concentration of mitochondria arranged in such a way that the form of energy which emerges is heat. This tissue produces additional heat to assist the infant in withstanding cold conditions during the early days of its life, when it isn't mobile and therefore not producing a great deal of heat through muscular movement.

So we find mitochondria in all living cells, providing the form of energy which those cells need. But how did they get there in the first place? There are some interesting ideas about their origin, as we will see next.

The origins of mitochondria

The singular of mitochondria is **mitochondrion**. Unlike the other sub-cellular structures, each mitochondrion contains its own genetic information in the form of DNA, and can divide to produce two new mitochondria independently of the division of the nucleus.

During sexual reproduction, the egg contains mitochondria derived from the mother. The sperm also contains mitochondria, as you saw in *Figure 14*, but these don't survive in the egg after fertilisation. This means that mitochondria are transmitted through the female line only. Some scientists believe that mitochondrial DNA represents a direct link back to the first human females.

The origins of mitochondria lie further back in time than that, however. Because they have their own DNA and can grow and divide independently, many scientists believe that they originated as free-living, independent bacteria. Somehow, these took up residence inside living cells early in the history of life on earth. The arrangement proved beneficial to both parties, so the bacteria remained in the cells and developed into mitochondria as we know them today.

The structure of mitochondria

Mitochondria are larger than many of the sub-cellular structures such as ribosomes. They are just about visible when a living cell is viewed with the ordinary light microscope. In different cell types they vary in shape, size and numbers. When they are studied at very high magnification using an electron microscope, it can be seen that each mitochondrion has an outer and inner membrane boundary.

The inner one is folded inwards to produce a large surface area, rather like irregular shelves, inside the mitochondrion. At very high magnification, the shelves (**cristae**) are seen to contain batteries of enzymes arranged in regular patterns. Between the shelves lies the mitochondrial matrix, in which a further range of enzymes is located. *Figure 15* shows the structure of a mitochondrion as seen in an electron micrograph.

When tissues are prepared for electron microscopy they are sliced very thinly indeed. So the photographs which we see represent the cut surfaces of the structures through which each slice passes. *Figure 15* shows how an oval shaped mitochondrion looks if it is cut through lengthways.

From this brief description it should be clear that the inside of a mitochondrion is very highly organised.

Like everything in the human body, it is adapted to perform a particular task. This degree of organisation is required for efficient energy conversion, as we will see after the following brief summary.

Summary

- Usually the cells are supplied with plenty of oxygen by the blood.
- Under these circumstances, pyruvic acid and the hydrogen produced as a result of glycolysis enter the mitochondria.
- Mitochondria are sub-cellular structures found in all living cells.
- They are most plentiful where cells are undertaking intensive work.
- Mitochondria can move and divide independently of the nucleus. They contain DNA.

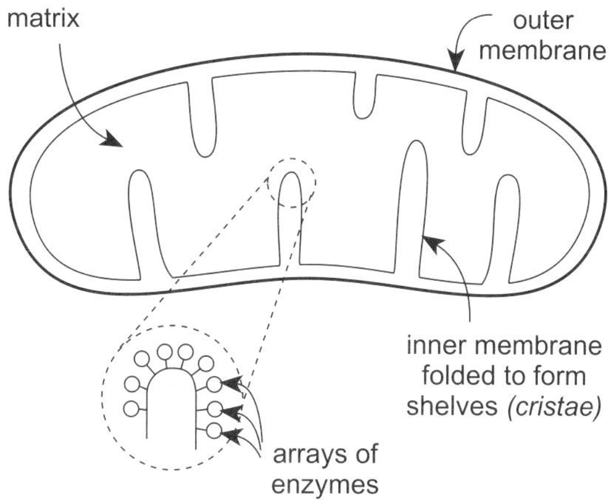

Figure 15 The structure of a mitochondrion

- For this reason, some scientists consider that they were originally free-living bacteria.
- Each mitochondrion is bounded by a double membrane. The inner one is folded into shelves (cristae) which produce a large internal surface area.
- The inside of the mitochondrion contains highly organised sets of enzymes.

What happens in the mitochondria?

Can you remember tracing the fate of glucose when it was being used as fuel in the absence of oxygen? We explained that this process gave a relatively small energy yield in the form of two molecules of ATP for each one of glucose.

The end products of this process, glycolysis, are pyruvic acid and hydrogen atoms. To liberate the energy which they contain in a useful form, they must enter the mitochondria.

They cross the mitochondrial membrane through specialised structures, which we could regard as the 'gates' to the mitochondrion in the same way that we have already visualised the 'glucose gates' into the cell. These gates only open if oxygen is present.

After pyruvic acid has passed through the gates, it undergoes a reaction to form a compound with just two carbon atoms. We'll call this the **C_2 acid**. The third carbon atom is converted into CO_2.

The C_2 acid first encounters a series of enzymes in the matrix. This series is called the **tricarboxylic acid cycle (TCA)** or sometimes the **Krebs' cycle**, after its discoverer.

This set of enzymes is found in all living cells, and represents one of the fundamental processes of life.

A brief outline of what happens goes something like this.

First, the C_2 acid combines with a C_4 acid to become a C_6 acid. This then undergoes a series of rearrangements, each catalysed by a different enzyme, in which:

- the original C_4 acid is reformed, ready to go round the cycle again

- the two carbons from the C_2 acid are converted to CO_2, and leave the cell as a waste product
- the energy from the chemical bonds in the C_2 acid is converted and handed on, with hydrogen atoms, to the final stage in the process which takes place on the cristae.

This process, together with glycolysis, is summarised in *Figure 16*.

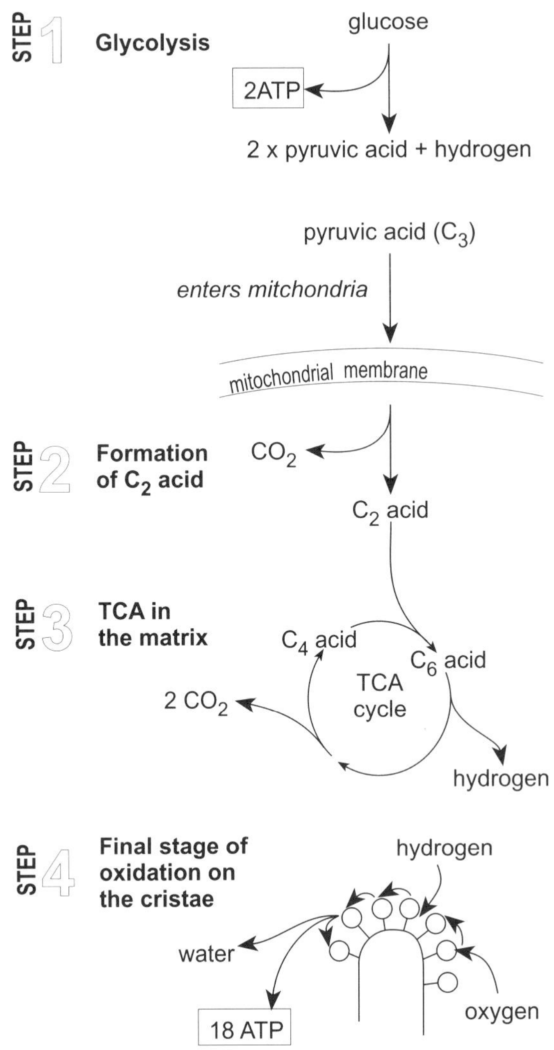

Figure 16 A summary of the complete breakdown of glucose

The final stage in the whole process occurs on the cristae when, through a further series of reactions, the hydrogen reacts with oxygen to produce water.

This is the stage when most of the energy is handed over to ADP, which has been formed during cellular work, and is ready to be recharged.

Eighteen ADP molecules are 'recharged' to ATP in the mitochondria for each molecule of pyruvic acid. However, each glucose molecule gives rise to two of pyruvic acid, so the total number of ATPs per glucose is 36 molecules through the TCA and two from glycolysis – a total of 38. It is not essential to remember these particular numbers, but it is important to notice the difference between the gain in ATP (two molecules) during glycolysis in the absence of oxygen, and the gain during complete oxidation of glucose through the TCA (38 molecules).

All the processes occurring in the mitochondria can be summarised by a third equation which shows what happens to the pyruvic acid. Instead of acting as a dump for the excess hydrogen, as in *Equation 2*, it is completely broken down to carbon dioxide.

Equation 3

$2 \times C_3H_4O_3$ + 4H + $6O_2$ + 34ADP +34P → $6CO_2 + 6H_2O$ + 34ATP
(pyruvic acid) (hydrogen) (enzymes of TCA)

ACTIVITY 35

ALLOW **10** MINUTES

This activity provides practice in answering equations.

First, check that *Equation 3* balances. Satisfy yourself that there are the same number of atoms on each side of the arrow. Are all the atoms which were present in the molecules of pyruvic acid accounted for?

Secondly, combine *Equations 1* and *3* together to give the overall summary equation for the breakdown of one glucose molecule. In other words, write down the two equations one after the other, and remove the intermediate stages.

For example, if our two equations looked like this:

A + B = C + D

and

C + D = X + Y

then writing the two together gives

A + B = C + D; C + D = X + Y

Therefore, we cancel the intermediate stage and say

A + B = X + Y

Your task is more complicated than this, but it is possible to combine the equations in a similar way, and discover that they still balance. It will be easier to do this if you leave out the ATP and write the equations like this.

Equation 1

$C_6H_{12}O_6$ → $2 \times C_3H_4O_3$ + 4H [Gain = 2ATP]
(glucose) (enzymes of glycolysis) (pyruvic acid)

Equation 3

$2 \times C_3H_4O_3$ + 4H + $6O_2$ → $6CO_2 + 6H_2O$ [Gain = 36 ATP]
(pyruvic acid) (oxygen) (enzymes of TCA)

Commentary

Writing *Equations 1* and *3* together gives

C6H12O6 → $2xC_3H_4O_3 + 4H$ [Gain = 2ATP]

$2xC_3H_4O_3 + 4H + 6O_2$ → $6CO_2 + 6H_2O$ [Gain = 36 ATP]

Equation 4

$C_6H_{12}O_6 + 6O_2 \rightarrow 6CO_2 + 6H_2O$ [Gain 38 ATP]

This last equation is the summary of the complete breakdown of glucose in the presence of oxygen – the complete oxidation of glucose.

This may already be familiar to you as it is often used, in simplified versions, to describe the fate of glucose inside living cells:

glucose + oxygen → carbon dioxide, water and energy

or $C_6H_{12}O_6 + 6O_2 \rightarrow 6CO_2 + 6H_2O$ + energy

I have unpacked this simple equation because I wanted you to appreciate that it is a summary of a very complex and beautifully organised sequence of events that is happening at this moment in every cell of your body, and on which all life depends.

You may prefer to use the simplified version to summarise the sequence.

There is no need to remember the details of the other one, but I hope that it has given you an appreciation of what is really involved in this apparently simple process, which is usually called **tissue respiration**.

Efficiency and pollution

The requirements for a real power station are that it should convert energy to a useful form efficiently, with a minimum of waste products and with as little harmful pollution as possible.

The cell has the same requirements. When oxygen is present, there is a good yield of ATP for every molecule of glucose fuel used. Energy which is not 'captured' as ATP is liberated as heat, which is useful to maintain body temperature, so the overall process is efficient.

The waste products are water and carbon dioxide. The latter is the only harmful source of pollution, and requires a mechanism for its removal from the body. The water is an important internal source for the body and is either reused or removed into the bloodstream.

Summary

- In the mitochondria, pyruvic acid is completely oxidised to CO_2 and water.
- This occurs in two complex series of reactions.
- The first is the TCA cycle, carried out by enzymes in the mitochondrial matrix.
- The second takes place on the mitochondrial cristae. Here, the hydrogen and energy 'bundles' react with oxygen. During this process, the energy is channelled into ATP production.
- A total of 38 molecules of ATP are made overall during the oxidation of one glucose molecule, including the two produced during glycolysis.
- The overall equation for the oxidation of glucose, tissue respiration, is $C_6H_{12}O_6 \rightarrow 6CO_2 + 6H_2O$.
- This system is efficient and the only harmful waste product is CO_2.
- ADP is recharged, re-forming ATP, on the mitochondrial cristae.

- This process only occurs in the presence of oxygen.
- The main function of the mitochondria is to provide a constant supply of the energy source ATP.

Alternative fuels

Although glucose is the main fuel source for cells, and is the only source that brain cells normally use, other foods can also be used to produce ATP and so supply useful energy.

You may recall that proteins are broken down into their building blocks, the amino acids. These are mostly required to build cellular proteins; therefore they are too valuable to be used as fuel – using them in this way would be like burning your furniture to keep your house warm. However, the body has no way of storing amino acids so, if there's an excess, some of them will be converted into glucose or into the acids that take part in the TCA and will be channelled into the oxidative pathway described earlier.

Fats are a true alternative fuel supply. Indeed, they are the preferred supply for some types of muscle. They provide the basic materials for the construction of many cellular structures. Excess fats can be stored and readily converted into one of the intermediates in the chain of reactions through which pyruvic acid is oxidised. They thus feed into the process which produces ATP in the mitochondria. I will describe exactly how this happens when we discuss the integration of these metabolic pathways in the final session of this unit.

Fats are a very good fuel source for most cells, but if they are used exclusively a pollution problem can arise. In order for fats to enter the TCA some carbohydrate fuel must be used as well. We could say that if fats are to be used as fuel in the cellular power stations, a carbohydrate 'booster' is needed. If this is not available, by-products from fat oxidation pile up. These are compounds called **ketones** and they result from the incomplete 'burning' of fat in the absence of carbohydrate.

Nail polish remover contains a member of the ketone family – **acetone**. The ketones produced in the body have the same kind of smell as acetone, and this smell can be detected on the breath if ketones accumulate. This is not a problem in itself, but accumulated ketones are acidic and they reduce the pH of the blood plasma. This affects many parts of the body, including the brain. If uncorrected, progressive acidosis leads to coma and death.

A healthy person uses a mixture of the two main fuels under normal conditions and therefore has no problem with the accumulation of ketones. However, if food intake is restricted for any reason, the body will start to use up its fat reserves and the proportion of fat used as fuel will increase. Ketones start to accumulate.

In small amounts, these can be removed by an increase in the output of urine – they are 'washed out' of the blood. This usually happens when a person starts to lose weight by dieting. They may be detected in the urine by an appropriate test.

A person with untreated diabetes is unable to use carbohydrates within his or her cells, and is in the same position as someone who is starving. Their bodies will use their fat stores as fuel and ketones will accumulate. Large quantities of urine are passed because the body needs to get rid of the ketones. This results in thirst and dehydration. The classic early symptoms of diabetes mellitus are thirst, weight loss and a high output of urine, as we saw in Jason's case study in *Activity 26*.

ACTIVITY 36

ALLOW 15 MINUTES

Returning to the case study of Jason, this activity asks you to apply all the biochemical understanding you have achieved so far to this case.

First, read the following case study and review the relevant sessions of this unit. You could also read *Resource 1* and *Resource 5*, which provide additional background material. As you read these sources, consider the underlying biochemical events which are giving rise to the observed symptoms. After you've read the case study, write a short account – just a single paragraph – of these events.

Six months after his doctor's provisional diagnosis was confirmed, Jason is brought into the Accident and Emergency Department late one Friday evening. He had collapsed in the street outside a city-centre pub.

On arrival it looks as though he is just another heavy drinker, but a routine blood test shows a very high blood sugar level. He is deeply unconscious. He looks very undernourished, his eyes are deeply sunk in their sockets and his skin looks strange. When pinched, it doesn't quickly spring back to normal.

Jason's breathing is deep and rapid. He has obviously been drinking as he smells of beer, but there is another fruity smell too, as if he'd been mixing beer and sherry in quantity.

The friend who accompanies him tells the staff that he is a diabetic – a fact which Jason has only recently discovered. Apparently he doesn't talk about it a lot, so his friend doesn't know how he deals with his injections.

Now please write your short account of the biochemical events occurring in Jason's body.

Commentary

Jason could well have been mistaken for a drinker, but clearly his condition is related to his diabetes. Since his blood sugar is so high, it looks as if he has been neglecting to take his insulin.

The high blood sugar shows that his cells are not taking in glucose. It is not passing through the gates in the cell membranes, as described in Session Four. His body is therefore using fat as its main energy source, and ketones are accumulating as a result, giving rise to the fruity smell on his breath.

As the ketones have accumulated, the pH of his blood has fallen, affecting his rate of respiration and causing progressive unconsciousness.

Jason's body will have produced large amounts of urine to remove ketones and glucose from his blood. He is therefore dehydrated, as shown by his sunken eyes and inelastic skin.

Jason is going to need rehydration and the right amount of insulin to restore his blood sugar to normal, and to enable him to use glucose as a fuel in his cells. He is also going to need education and support when he is over this crisis, to help him to adjust to a problem which will be with him for life.

The episode we have just described is a **diabetic coma**, caused by a high concentration of sugar and ketones in the blood. Jason will have to learn to cope with the diabetic's other problem – too little blood sugar caused by an imbalance between insulin injected and carbohydrate in the diet. You can read how this affects a person in *Resource 5*.

Summary

- Both fats and proteins can be oxidised through the same metabolic pathway as carbohydrate.
- Since proteins are needed for other purposes, they are not normally used as a fuel supply.
- Fats are a major alternative to glucose as fuel supply.
- Oxidation of fat requires a carbohydrate 'booster'.
- If this is not available, fat oxidation results in the accumulation of harmful ketones.
- Diabetes mellitus results from a failure of insulin production.
- If insulin is not replaced in people with this condition, fat is oxidised, ketones accumulate and severe symptoms follow.

This concludes our study of the way that cells use fuels, derived from food, to produce useful energy for their work.

The hormone insulin has figured in this session, and will do so again in Session Six as we consider how proteins such as this are packaged by the cell for transport to the place where they will be used.

Before you move on to Session Six, check that you have achieved the objectives given at the beginning of this session and, if not, review the appropriate sections.

SESSION SIX

Packaging and transport

Introduction

We have now covered most of the import and manufacturing side of the cellular small business, as we have considered the intake and use of raw materials and fuels. We can now go on to see how the products of the cell are exported to wherever they are to be used in the body.

This session takes a brief look at the way that cells package and label their products and sort them for despatch to different destinations. Our understanding of this kind of cellular work is in its early stages at present, and the state of knowledge about many of the details is changing. This will, therefore, be a short session, but this doesn't mean that the topic is less important than others in this unit.

The session concludes by returning to one of the specialised cells that we have used as an example throughout this unit – the β cell of the pancreas – and considers the way that it packages and despatches insulin. This example will illustrate some of the problems which have to be overcome by cells producing materials for export.

Session objectives

When you have completed this session, you should be able to:

- describe how materials get in and out of the different parts of the cell
- explain how proteins are packaged for transport when they are complete
- outline how manufactured products get out of the cell.

1: A second look at protein manufacture

In Session Two, I described how a protein was synthesised on the ribosomes, using messenger RNA from the nucleus as a 'pattern'. You may want to look back at this session, to refresh your memory.

We saw in Session Five that all the work the cell does, including synthesis, needs a great deal of energy. In fact, each time two amino acids are joined together, seven ATP molecules are used up. A single protein molecule may contain hundreds of amino acids, and will therefore demand a large input of ATP. Since production of ATP depends on a good fuel supply, protein manufacture depends on the availability of carbohydrate and fat.

This is the first example of the **interdependence** of biochemical processes which we have encountered in this unit. As we shall see, no cellular activity takes place in isolation. All of the many series of enzyme reactions are interlinked.

Proteins are manufactured on the ribosomes. Some of them will be made for immediate use in the cell itself. These are formed on the free ribosomes – those that are scattered around throughout the cell. For instance, the ten enzymes that bring about the reactions of glycolysis are made in this way.

Other proteins are made to be exported from the cell or to be stored for later use. These are made on ribosomes which become attached to the endoplasmic reticulum.

I will now explain why this difference exists and why some proteins have to be carefully 'packed' for transport through and out of the cell where they are made.

2: Packaging and sorting proteins

Endoplasmic reticulum

Any material which the cell makes for immediate use can be liberated from its place of manufacture directly into the cell. However, some of the cell's products need to be transported away from the production line.

During this journey, they could meet unfavourable conditions and be damaged, or they might themselves do damage to parts of the cells through which they pass. For example, digestive enzymes which break down proteins are made inside cells in the lining of the intestine and exported straight into the digestive system. On the journey, they encounter many of the cell's own proteins. They have to be 'packaged' in such a way as to prevent them damaging the actual cell where they are produced. Therefore they cannot simply be liberated from free ribosomes, like the glycolysis enzymes. They must somehow be kept separate from the other contents of the cell. This is where the endoplasmic reticulum comes in.

The endoplasmic reticulum (ER) consists of a network of tubes running through the cell. Newly-made proteins can be sent directly into these tubes. In this way, they are safely channelled away from the other contents of the cell.

Ribosomes which make proteins to be sent out of the cell, or which need to be kept separate for any other reason, are therefore located on the ER. When a protein being made on these ribosomes is complete, it doesn't separate from the ribosomes and float free in the cell. Rather, it immediately crosses the membrane into the ER.

Once in the ER network, the proteins are modified in a number of different ways. For example, most proteins destined for export out of the cell where they were made have a carbohydrate molecule added on to them. This appears to act as a 'label' which determines where they will be sent.

The modified proteins are then channelled into small swellings of the ER, known as **transport vesicles**. These become detached from the ER and carry the proteins through the cell to the next stage in the process – the packaging and despatch department – which is found in a structure called the **Golgi complex (GC)**, which is close to the part of the ER from which the transport vesicles arise.

In this way, the proteins remain protected from the other contents of the cell as they are transported from the ER to the Golgi complex.

The Golgi complex

This is another network of tubes, similar to the ER but without ribosomes. It has an irregular shape but usually, in electron microscope pictures, it clearly has two distinct sides, one more curved than the other. One side is always near the ER, and the transport vesicles arrive here, transferring the newly-made proteins into the network of the Golgi complex.

The Golgi complex contains a set of enzymes which deal with all the modifications that need to be made to the proteins before they can be sent on. The proteins are then transported through the GC and, in a way that is not yet understood, they are sorted out according to their 'labels'. They are then placed in the appropriate container for onward transport and leave the GC at the opposite side from where they started. This is shown in *Figure 17*.

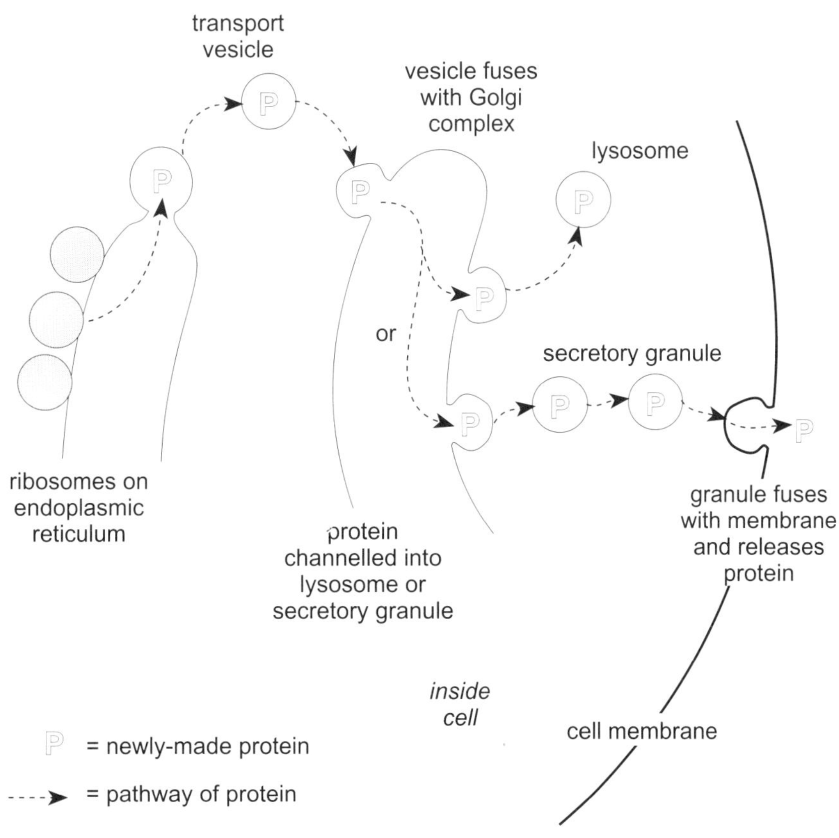

Figure 17 The relationship between the endoplasmic reticulum and the Golgi complex

Summary

- Protein manufacture is dependent on the availability of fat and carbohydrate fuels.
- Proteins for immediate use in the cell are made on free ribosomes. Those for export or storage are made on ribosomes attached to the endoplasmic reticulum.
- Newly-made proteins are transported from the endoplasmic reticulum direct to the Golgi complex.
- In the Golgi complex, they are sorted and placed in the appropriate container for onward transport to wherever they are needed.

3: Exporting proteins

Onward transport

The 'containers' which leave the export side of the Golgi complex include **lysosomes** and **secretory granules**, both of which are shown in *Figure 17*.

Lysosomes transport very powerful enzymes, which are capable of destroying cellular macromolecules of all kinds.

You might wonder why a cell would contain such a destructive mixture. A clue to their use is found in the fact that lysosomes are particularly abundant in white blood cells, which act as the first line of defence against bacterial infection. Hence, they take in and destroy foreign objects, such as bacteria.

In addition to this, the lysosomes recycle cell contents. They take in worn-out structures, such as mitochondria, and dismantle them. The raw materials are then released into the cell for reuse, whilst any waste materials remain in the lysosome, which now acts as a cellular dustbin.

Some lysosomes also have tasks to do outside the cells in which they are made. For instance, the cells which reshape bones after injury produce lysosomal enzymes which digest the bone itself. These lysosomes take the powerful enzymes out of the cells and release them where they are needed.

If a lysosomal enzyme is defective, the substance which it normally destroys will accumulate in the lysosomes, and may produce serious symptoms.

This occurs in several uncommon serious or even fatal genetic diseases, showing how important these structures are in removing harmful materials from the cell.

The interior of the lysosome is more acidic than the rest of the cell, and its enzymes work best at the lower pH. This is an additional way in which the cell itself is protected from the lysosomal enzymes.

Lysosomes, then, are the containers which convey a range of destructive enzymes to their site of action, or keep them ready for use.

Secretory granules represent another type of onward transport – this time taking newly-made proteins, such as insulin, out of the cell. We will see how this occurs next .

Summary

- Lysosomes contain powerful destructive enzymes which are kept separate from the contents of the cell.
- Lysosomes have three functions: to destroy invading bacteria; to recycle worn-out parts of the cell; and to carry powerful enzymes out of the cell for particular tasks, such as bone reshaping.
- Genetic defects in lysosomal enzymes may produce serious symptoms.

Out of the cell

Many of the products made by specialised cells have to be ready at a moment's notice. In other words, they must be immediately available as soon as they are needed. Substances of this kind are made in advance and stored until they are required.

Cells are very efficient in the way that they use space. Substances being stored are usually packed as tightly as possible into the available space. Secretory granules contain the products which they are storing in concentrated form, ready to be released when required.

Cells which specialise in exporting their products often show a well-organised internal structure in which raw materials come in at one end and the products are stored and secreted at the other. *Figure 18* shows a cell of this kind in action.

ACTIVITY 37 — ALLOW 5 MINUTES

This activity will provide you with useful revision of the topics covered in the last two sessions.

Figure 18 shows a cell which secretes digestive enzymes. The timescale for its activity is also shown.

Study the diagram and identify the stages in the manufacture and export of its product, starting from the bottom of the cell, labelled '1', and working up to the top.

At each place numbered on the diagram, write a brief sentence to describe what is happening.

1

2

3

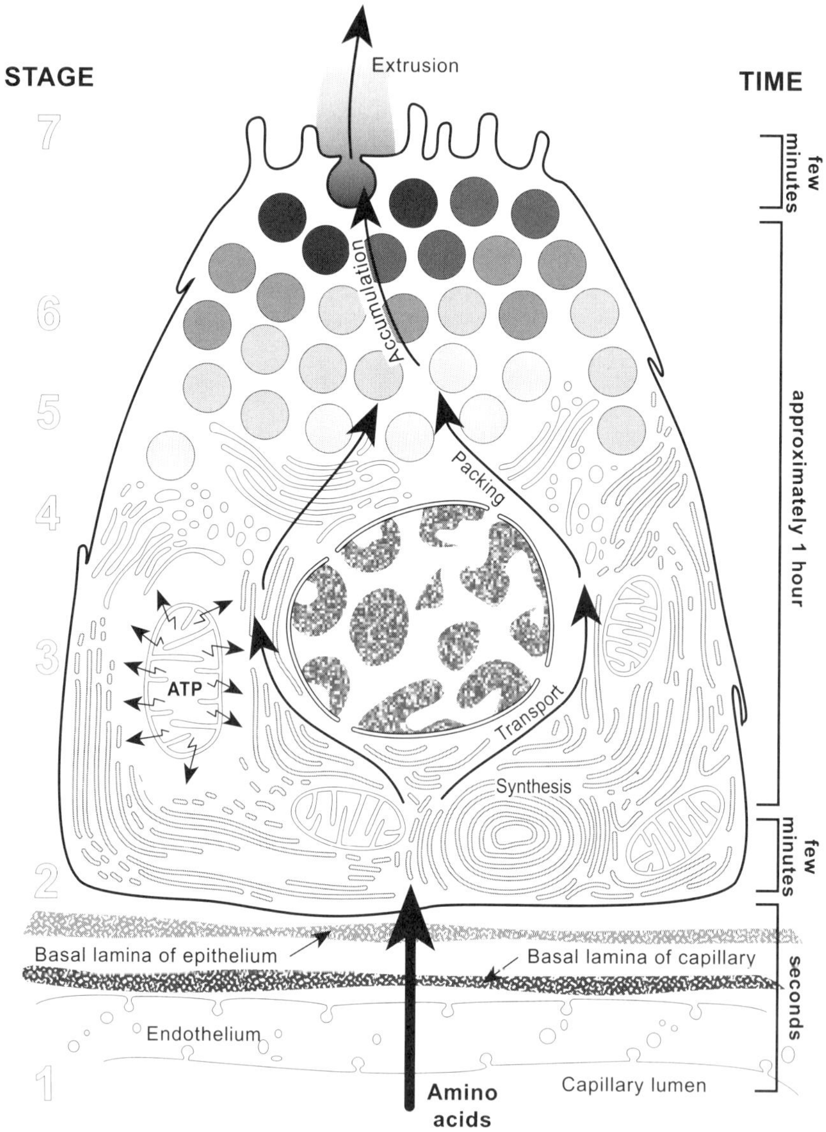

Figure 18 An enzyme-producing cell

Source: Adapted from Junqueira, Carneiro and Kelly, Basic Histology, 7th edn, Large Medical Books, Appleton and Large, Prentice Hall International, 1992, p.84, Figure 4-19

4

5

6

7

Commentary

Here is the sequence of stages in *Figure 18*. The wording of your answers may differ from mine, but these are the key points you should have covered.

1 First, amino acids enter the cell from the blood in the small blood vessel – the capillary.

2 Next, the amino acids are joined up to form proteins on the endoplasmic reticulum.

3 The mitochondria provide energy for this process in the form of ATP.

4 The new proteins are then sent to the Golgi complex.

5 The proteins are packaged in the Golgi complex.

6 The secretory granules accumulate at the top of the cell.

7 Finally, the product is released as required.

I hope that this exercise has refreshed your memory about the whole process. Now let's take a look at a specific example of protein packaging and transport.

Insulin production

Insulin is a small protein, produced in the β cells of the pancreas. It acts as a hormone by combining with receptors on the surface of cells and encouraging the uptake of glucose. As we said in Session Four, it 'opens the glucose gates'.

Insulin is synthesised in a similar way to the process you studied in *Activity 36*. A message from the nucleus, in the form of messenger RNA, signals the start of insulin production. The ribosomal activity constructs an inactive form of the hormone, pro-insulin, which is longer than insulin itself. This is transported to the Golgi complex, where it is converted to insulin by an enzyme which snips off the masking amino acid chain.

The active insulin is then packed into secretory granules, which are transferred to the edge of the β cell. When blood glucose reaches a certain level, a signal is sent to the β cells and insulin is discharged into the blood, in the small capillaries which are close to the β cells. From there, it reaches the general circulation.

Energy is needed at many stages in the process of producing insulin. It is needed to transport the amino acids into the cells, to join them together, and to discharge the insulin from the cell. As you might expect, insulin-producing cells are well supplied with mitochondria, which produce the ATP to supply all the energy required.

This example sums up all the stages in protein production, and links them with the production and use of ATP. You may find it helpful to turn back to *Table 1* now and fill in further adaptations of the pancreas β cell that you can now identify.

We have now almost covered the full range of activities of the cellular small business enterprise. In the final session, we will deal with the last two vital functions – the integration and control of all these activities.

Summary

- Many cells produce substances which are exported – or secreted – from the cell.
- These substances are made in advance and stored in secretory granules until required.
- The substances are secreted when the cell receives a signal that they are needed.
- Insulin is first made in its inactive form, and then stored in secretory granules.
- Insulin is released into the bloodstream when blood sugar rises above a certain level.

Before you move on to Session Seven, check that you have achieved the objectives given at the beginning of this session and, if not, review the appropriate sections.

Session Seven

Quality control

Introduction

To end this unit, I would like to discuss a function which is of increasing interest to health professionals – quality control. This is becoming a discipline in its own right in the health services, but it is no new development as far as the cell is concerned. Words like 'monitoring' and 'standard setting' have always been applicable to cellular activities.

In this concluding session we will briefly consider the functions of control and integration as they apply to a small business. Then we'll apply these ideas to the work of the cell, dealing with the role of hormones as controllers and focusing again on insulin. Finally, we will develop the idea of integration of all the cellular processes to produce quality output.

Session objectives

When you have completed this session, you should be able to:

- explain the need for control and integration within the cell
- describe methods of internal and external control of cell metabolism
- outline the role of hormones in external and internal control
- describe the action of insulin
- explain the concept of integration as applied to metabolic pathways in the cell.

1: Keeping control of a small business

The analogy used throughout this unit is that of a small manufacturing company which carries out a wide range of different kinds of work.

Clearly an organisation such as this needs to have someone in control to ensure that the right products are made at the right time and in the right quantity. This means that the company must be able to detect and respond to the needs of its market, that the speed of its various production lines must be regulated, and all its activities co-ordinated. Furthermore, the quality of its products must be checked and steps taken to ensure that any decline in quality is remedied before it gets out of hand.

All these tasks are carried out by the management team. Each living cell has to be subject to similar processes of control and integration but it has no management team. Its organisation is self-regulating.

Biochemists have worked out the sequence of events in the cell's metabolism and now have a good understanding of the nature and function of enzymes. They have discovered the details of many of the structures inside cells. But they are only just beginning to gain a full appreciation of the cell's internal quality control processes. We are going to touch on some of these now.

2: Methods of controlling cellular activity

The points of control

Since most of the cell's work is carried out by enzymes, we might expect that control of cellular activity will involve control of the cellular machinery. We also know that the activity of the machinery is determined by instructions from the nucleus, so this seems likely as another control site.

Both of these ideas are correct. The control may be done from either inside or outside the cell itself.

Internal control

The self-regulating capacity of the nucleus is a complex subject which is largely outside the scope of this unit.

I will just touch on it by mentioning one interesting aspect – how 'mistakes' are corrected. If certain kinds of damage to the DNA occur, enzymes in the nucleus are able to cut out the damaged piece and make good the defect. This implies that the cellular machinery is able to detect a faulty product – the damaged DNA – and take steps to improve it. In other words, there is a standard for DNA production, and a monitoring process to ensure that this standard is achieved.

This is a process of fundamental importance which protects the genetic material, and therefore the cell itself, from many naturally occurring hazards and accidents.

It ensures that, as far as possible, the 'quality' of the DNA is maintained.

That is all we need to say about the internal regulation activities of the nucleus. Let's now consider the regulation of enzyme activity. If you think of enzymes as pieces of machinery again, it might help us to develop a model of regulation.

ACTIVITY 38

ALLOW 2-3 MINUTES

This activity will help you to understand the how the activity of enzymes is regulated.

Imagine that you are in charge of a factory full of machinery producing a particular product. If you need to vary the output, you could do it in a number of different ways.

Can you suggest two things you could do to the machinery which would enable output to be varied?

1

2

Commentary

Perhaps the first thing that springs to mind here is to alter the rate at which the machines are working. You could speed them up or slow them down, and so alter their output.

Another possibility is to alter the number of machines in operation. You could take out some of the production line or, if business was good, you could buy in more machines to increase output.

These are the options open to the cell, too. In order to regulate its output of any product, the rate at which enzymes work or the number of enzyme molecules present can be changed. Let's look at each of these things in turn.

Very few enzymes act in isolation. They all take part in sequences where the product of one enzyme becomes the substrate of another. We have already seen two such sequences – glycolysis and the tricarboxylic acid cycle – in Session Five.

The speed at which any sequence takes place will depend on the speed of the slowest enzyme. This enzyme will control the speed of the whole sequence, and is known as the **rate-controlling enzyme**. In order to affect the rate of the whole sequence, it is only necessary to influence the speed of this particular enzyme.

So how is this done? Our model of regulation from *Activity 38* will help us to provide the answer.

ACTIVITY 39

ALLOW 2-3 MINUTES

Returning to the idea of the factory from *Activity 38*, imagine again that you are in charge of the production of a perishable product – let's say, trifles. What piece of information would make you decide to slow down the production line?

Commentary

There are a number of alternative answers here. As the manager of trifle production, you would no doubt receive continual reports of the state of the market and you might use those as a basis for forward planning. However, on a day-to-day basis, you would also be receiving information from the refrigerated stores where the trifles are held before despatch. If these stocks rose above a certain level, you would be likely to slow down production.

In the cellular factory, the same kind of control takes place. If the end product of an enzyme sequence starts to accumulate, the rate-controlling step in its production will be automatically slowed down, and production will also slow down until the surplus is removed.

This is termed **feedback inhibition.**

You may recall that glycolysis is a sequence of ten reactions in which glucose is split into two molecules of pyruvic acid and ATP is produced. We could represent this sequence as:

Glucose →	A →	B →	Pyruvic acid
Enzyme 1	*Enzyme 2*	*8 further steps* *Enzymes 3–10*	

Enzyme 2 is actually the rate-controlling enzyme, and it is *slowed down by* ATP. So the more ATP that accumulates, the more slowly glucose is used.

ACTIVITY 40

ALLOW 2-3 MINUTES

This activity will help you to understand the concept of feedback inhibition.

Suppose that glycolysis is taking place under anaerobic conditions in a particular tissue. That tissue is not very active at this time. It has a good supply of glucose. What do you think will happen? Use your knowledge of biochemistry to predict what might happen to the glucose in that tissue, and sum it up in a single sentence.

Commentary

You might have said that:

'Glucose will be built up into glycogen for storage'.

If the tissue is not very active, it won't be doing a great deal of work. It won't be using a lot of energy. Therefore, it won't be using the ATP made during glycolysis. This will build up and will slow down the rate-controlling enzyme in the glycolysis pathway. This, in turn, slows down the whole process, so glucose won't be used up. There will be a surplus of glucose, which means that it will be converted into glycogen.

This last part of the chain of reasoning would have involved a little guesswork on your part, as we haven't said anything about glycogen production yet. In fact, the synthesis of glycogen from glucose is not a long sequence – it only involves two enzymes. One of these (**glycogen synthetase**) brings about the build up of glycogen and the other (**phosphorylase**) catalyses its breakdown. We could represent this as:

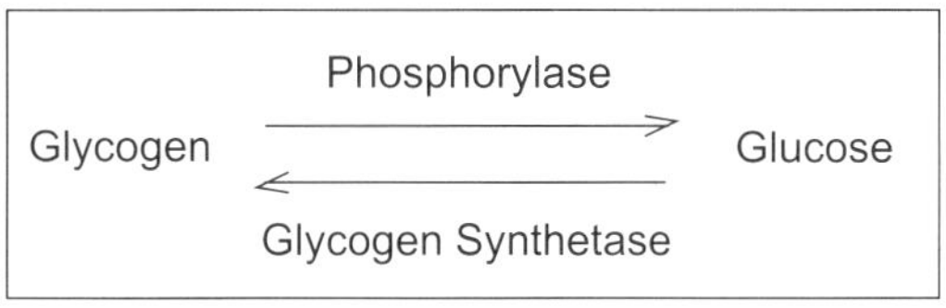

Each of these enzymes exist in an active and an inactive form. They can be switched from one form to another by several agents – including ATP and adrenalin. The switching agents have opposite effects on these enzymes – they switch glycogen synthetase off when phosphorylase is switched on, and vice versa. Thus, if there is plenty of ATP in the cell, the enzyme making glycogen is switched on and the enzyme converting glycogen to glucose is switched off.

The switch mechanism is not an all or nothing effect. The activity of each of these enzymes can be regulated in a series of small steps, so the overall control can be finely adjusted to meet the cell's demands.

So the presence of this end product of glycolysis, ATP, does two things. It slows down glycolysis so that there is no over-production, and it diverts the raw material, glucose, into storage, so that there is no waste. This is therefore a very neat mechanism for regulating glucose use in glycolysis.

Every one of the cell's enzyme sequences is internally controlled in similar ways, by adjusting the rates and directions in which the enzymes work. This kind of control happens very rapidly, so the enzyme sequences are very responsive to local conditions.

The result of this internal control system is that the amounts of all the chemicals in the cell remain constant. If any key substance, such as ATP, accumulates in the cell, the internal control system will act to slow down its production and speed up its removal, thus returning the level to normal and maintaining constant conditions inside the cell.

The other control method suggested in *Activity 38* was to change the number of machines in operation, or, in the cell, to vary the number of active enzyme molecules present. This is done by varying the rate at which enzymes are broken down (probably by the enzymes in the lysosomes), or the rate at which they are produced, by varying the rate of protein synthesis.

This second kind of control is a longer term measure. It does not happen as quickly as the previous control measure, but represents a response to a persistent change in the cellular environment. This may be the result of a change in the person's food supply or the administration of a new drug over a period of time. The enzymic machinery can respond by producing more enzymes to deal with a new challenge.

For example, when a new drug is administered to a person, its presence in the body constitutes a change in the cellular environment. The drug circulates in the bloodstream and is distributed from there into the fluid which surrounds all the cells. If the drug is given by mouth, it will first pass through the liver, because all the blood from the digestive system is routed through the liver on its way to the general circulation. The liver contains many enzymes whose task is to deal with any foreign substances. In prolonged or extensive drug usage, the liver increases production of these protective enzymes.

Summary

- Cells need a quality control system to ensure that they produce the right thing at the right time.
- They therefore need control and integration systems.
- These may be internal or external to the cell.
- Internal control is brought about by regulation of the activity of the nucleus, or by varying the rate of action of enzymes in the cell.
- In an enzymic system, there is usually a rate-controlling step. Regulation occurs at this point.
- The regulation may result from feedback inhibition or from altering the enzyme's state.
- This type of regulation provides a rapid response to local changes.
- A slower response to environmental change occurs when the amount of a given enzyme increases to deal with a new challenge.

Having outlined some of the ways in which the activities of the cell are regulated from within, I would like to turn to the question of external regulation – how agents produced in other parts of the body affect the way that cells operate.

External control

Returning to the model used in *Activities 37* and *38*, we might begin to wonder what other factors beside the accumulation of product could influence the rate of production in the factory. Experience suggests that the manager would have to plan production in response to directives from Head Office, and not just in response to the state of the storage depot. The production of trifles might only be one component in a vast enterprise, and its rate would be tied in with that of all the other products.

In just the same way, no individual cell can act in isolation in the human body. It influences, and is influenced by, the activities of other cells. A mechanism for the co-ordination and control of all tissue cells is needed, and this is what regulates cells from 'outside'.

This mechanism consists of two parts:

- nervous control
- hormonal control.

It brings about a situation in which the internal conditions in the body are constant, so that the enzymic machinery in all the cells can operate as smoothly as possible.

This process of maintaining constant internal conditions is called **homeostasis**. The term covers all the processes which work together to keep the intercellular fluid (the fluid which surrounds all cells) at the temperature and chemical composition which best suits the cells.

It is relatively easy to observe some homeostatic processes in the body as a whole. For example, if a person becomes very hot, their body automatically makes adjustments to lose heat by sweating. Physiology is the study of homeostatic processes at this level.

Underlying these processes are mechanisms which operate at the biochemical level – in and around the cells. We are now going to consider how nervous and hormonal controls operate at this level. This should help to show you how physiology and biochemistry are very closely related. Let's begin with nervous control.

Nervous control

Some tissues are not under direct control by nerves. In other words, they do not have direct contact with nerves. For example, pancreatic cells are not directly influenced by messages from the nervous system.

Muscles, on the other hand, have 'direct lines' to the nervous system and their activities are controlled and integrated by messages which travel along nerves – nerve impulses.

Let us return to the specialised muscle fibres first shown in *Figure 1* and *Figure 7*. We can add a little more detail now, to describe how the activity of these cells is controlled by nerves.

As I mentioned earlier, the contractile elements in the muscle fibre consist of proteins. There are two, called **actin** and **myosin**. They are arranged in overlapping bands in the myofibrils.

Each fibre is supplied by a branch of a nerve. The nerve and muscle forms a single unit, as shown in *Figure 19*.

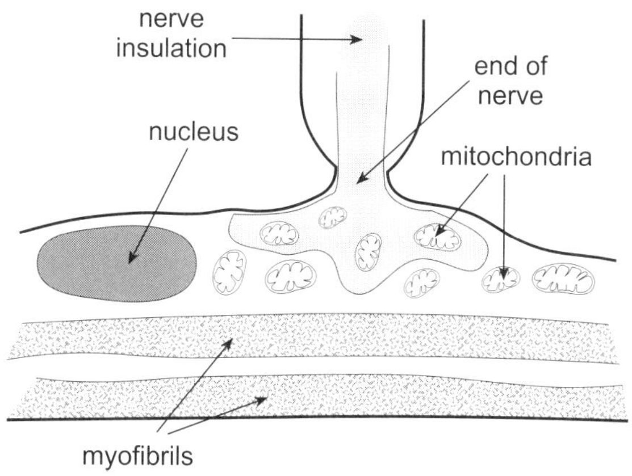

Figure 19 The junction between a nerve and a muscle fibre

When the body requires a particular muscle to contract, an electrical impulse is sent down the appropriate nerve to the place where it contacts the muscle. This causes the release of a substance called **acetyl choline** from the nerve ending, which penetrates into the fibre.

The effect of the presence of acetyl choline is to release **calcium** into the muscle. This triggers the contraction process. Actin and myosin are highly specialised proteins with the ability to form and break connections between themselves very rapidly, using ATP as the energy source. Once formed, each bond bends, pulling the two proteins over each other and shortening the fibre.

In the complete absence of ATP, the bonds become permanently fixed. This gives rise to **rigor mortis**, the rigidity which occurs after death.

The relationship between nerve and muscle is therefore a very close one. As we shall see, it must be maintained if the muscle fibres are to remain functional.

ACTIVITY 41

ALLOW 2-3 MINUTES

This activity will help you to understand the relationship between nerve and muscle.

People are very different in terms of the amount of muscle tissue they possess and, in the individual, the size and strength of muscles can change over time. Can you identify some of the causes of either an increase or a decrease in the amount of muscle tissue?

Commentary

You may have observed that exercise can increase muscle mass and perhaps, from your own experience, you might be aware that it also increases muscle strength.

Conversely, inactivity reduces muscle mass and strength. This is felt when a person gets up after being in bed for only a few days. It is seen to a greater extent in some people who have suffered damage to part of their nervous system. If the nerves supplying a particular set of muscles are severed by trauma, those muscles become limp and may **atrophy** – that is, they may waste away.

Normally, all our muscles are in a state of light contraction – they have a certain degree of **tone**, maintaining normal posture and ready to respond quickly when necessary. If the nerve supply to a muscle is destroyed, the junction between nerve and muscle degenerates, the muscle is incapable of contraction and has no tone at all. It is said to be **flaccid**. In time, the muscle fibres shrink and their structure breaks down.

These effects demonstrate that muscles must receive regular stimulation from nerves if they are to remain intact and functioning.

The nerves which supply all parts of the body have their origins in the central nervous system (CNS) – the brain and spinal cord. Their activities are co-ordinated by messages which travel through the CNS. Damage may occur in the brain – as the result of a stroke, for example. This interrupts the messages in the CNS, but leaves the nerves supplying the muscles intact. If alternative nerve pathways can be established muscles can still receive stimulation and they do not atrophy.

In severe accidents, where the spinal cord is severed, the relationship between nerve and muscle is clearly demonstrated. Injuries like this sometimes happen to young people during sports such as rugby and diving. At first, the person is completely collapsed and floppy. The muscles controlled by the part of the spinal cord below the break are completely flaccid.

After about three weeks some tone returns to these muscles and they respond to some kinds of stimulation, although they remain paralysed. In other words, some **reflexes** return but these are different from normal reflexes because they involve the spinal cord only, below the level of damage, and are not mediated by messages to and from the brain.

In a condition such as this, the first phase which is termed **spinal shock** shows the effect of the absence of nerve stimulation on the muscles. As this wears off, and new nerve pathways take over, we see the effect of abnormal patterns of stimulation. The way that people in this situation are positioned, handled and moved interacts with these patterns and is therefore of the utmost importance in helping them to regain as much normal function as possible. Nurses have a very important role here in the care of people who have suffered damage to the brain or spinal cord.

ACTIVITY 42

Here is a piece of investigation for you to carry out on your own or with a group of colleagues. This activity will help you to relate the work you have done so far in this session to the nursing care of someone who has suffered a stroke.

A stroke is a very common cause of illness in older people. It occurs, more rarely, in younger people too. Damage is caused to part of the brain by leakage of blood – hence its official name, **cerebrovascular accident (CVA)**.

1 From either observation or discussion with experienced colleagues, identify what happens to the limb muscles of a person who has suffered a CVA. You could choose a specific individual as a case study, if you wish.

2 Discuss the nursing care of such a person, to identify what is being done to ensure that he or she regains as much normal movement as possible.

3 See if you can provide an explanation at the biochemical level of what happens to the person's limb muscles during the first few days after the CVA.

Commentary

One thing that we can say for certain about people with CVAs is that the outcome is very variable, so your description may be very different from mine. However, our descriptions probably share some of these features.

1 The person was very floppy on one side at first. Then gradually the affected muscles developed greater tone, causing the affected limbs to bend and turn inwards.

2 The person was positioned carefully in bed at first to avoid the development of contractures. The limbs were exercised by passive movements.

As the person recovered, a rehabilitation programme was established to make use of the remaining pattern of movement in the affected limbs. (There is some debate about the best way to do this, so we will not go into detail here.)

3 Your description of the biochemical events might go something like this:

At first, the nerves supplying the affected muscles are not transmitting any impulses. No acetyl choline is being produced at the neuromuscular junction and no muscular contraction, involving actin and myosin, takes place.

During recovery, alternative reflex pathways develop and the muscles receive some stimulation. They do not atrophy but may show abnormal reflex activity and become fixed in abnormal positions, causing disability.

The care of stroke patients is another highly specialised area of nursing where an understanding of the biochemistry and physiology of nerve and muscle is needed for effective practice.

This exercise gives a little insight into the way that nerves directly control and integrate the activities of one type of cell – the limb muscle cell.

In fact there are different types of muscle in the body, each one adapted for a different purpose and differing in its relationship with the nervous system. For example, the muscles which control the eye movements are responsible for very small, controlled and precise movements. Here each nerve fibre is in contact with and controls just five muscle fibres, whereas in the large leg muscles, where movements are on a much larger scale, one nerve fibre can control up to 2,000 muscle fibres.

The muscles we have been discussing are called the **skeletal muscles**. Other muscles control the automatic functions of the body, such as the heart rate and activities of the stomach. These are controlled in slightly different ways. In all cases, response to nerve impulses is quick, demonstrating why this type of external control of cellular activity might be called 'rapid response' as it enables the muscle cells in all parts of the body to make adjustments for moment-by-moment changes in the environment.

Let's summarise the main points covered so far before moving on to discuss the second external control mechanism.

Summary

- External control of cellular activity is brought about by the nervous and hormonal systems.
- The end result of these control processes is homeostasis – the maintenance of a constant internal environment in which all cellular enzymes can work most effectively.
- The contractile elements of muscle are composed of two proteins, actin and myosin.
- Nerve fibres transmit electrical impulses to muscle and cause the release of acetyl choline, which in turn liberates calcium in the muscle, setting off the contraction.
- Actin and myosin form and re-form bonds using ATP energy. The bonds pull the threads of these two proteins together, thus contracting the muscle.
- Inactivity of the muscle results in loss of muscle mass and strength.
- The response of muscle to nervous control is rapid.

3: Hormones as external controllers

Hormones are substances produced in one part of the body, and having an effect in another – their target. Every hormone affects one or more specific target tissues.

In some cases, the hormone gets inside the tissue cell and works on the nucleus to control its activities. This is the first kind of hormone action which we will briefly consider.

Hormones in the nucleus

There are two kinds of hormone which we know enter the cell and affect the nucleus – the **thyroid** and the **steroid** hormones. I will use the second group as an example to show how they operate.

Steroids are relatively small non-protein molecules. They are known to pass easily through cell membranes and to become located in the nucleus of their specific target cells.They fulfil a variety of functions, but all have something in common.

ACTIVITY 43 ALLOW 2-3 MINUTES

This activity asks you to think about the functions of different kinds of steroids.

You may have read about 'anabolic steroids' in the press, particularly in relation to sports people. Other steroids you may have heard of include **testosterone**, the male sex hormone, and **progesterone** and **oestrogen**, the female sex hormones which regulate the menstrual cycle and pregnancy.

On the basis of your existing knowledge of these hormones, can you detect any similarity between the functions they carry out in the body?

Commentary

You may have concluded that all these hormones are concerned in some way with growth and development – long-term processes.

Testosterone, for example, is responsible for the development of male characteristics in the unborn child and in the appearance of secondary sexual characteristics at puberty.

These hormones exert a kind of long-term control which is responsible for the fundamental changes which take place at different stages in our lives. In order to do this, they have to get right to the heart of our cellular small businesses – right into the Planning Department, in fact.

They exert their control by entering the nucleus and influencing the action of the genes. For example, they sometimes switch on genes which have previously been inactive, and so cause a particular tissue to produce a new product. If you study the physiology of milk production (lactation), you will be observing the effect of such a profound biochemical change within the female body.

Because hormonal influences are so profound, times of hormonal fluctuation represent a considerable biochemical upheaval. We might even liken this to a time of organisational change in the cellular small business – with all the stress which that can include. Times of major hormonal change in our patients and clients represent a particular challenge for health-care professionals.

There is a great deal more which could be said about the role of steroid hormones, but now we need to move on to consider the action of the other main group of hormones – those that do not enter their target cell.

Hormones on the surface

When I discussed the way that substances get into cells in Session Four, I mentioned that cell membranes have binding sites for insulin.

This is also true for all the hormones which, like insulin, are small proteins. They are much bigger molecules than those, like steroids, which get right into the cells, and they remain on the surface of their target cells.

So we might ask, how do they exert any influence inside the cells?

One of the ways has already been suggested – they alter the rate at which raw material passes into the cells and therefore increase the rate at which the raw material can be used.

This is undoubtedly one of the ways that insulin works but, as in the case of all hormones, it is only part of the story. Insulin has a number of different effects, all of them related to the efficient use of fluctuating food supplies. *Table 5* lists some of these actions.

Insulin encourages:

- the uptake of glucose by cells
- the oxidation of glucose in cells
- the production of glycogen from glucose
- the uptake and use of amino acids by cells.

Insulin discourages:

- the breakdown of fat to provide energy.

Table 5 Actions of insulin

When the production of insulin increases, all the methods of using glucose are encouraged, and the uptake and use of amino acids is encouraged. This happens in response to a high blood glucose level – that is, when the body has a good supply of raw materials, such as when a meal has been eaten.

At the same time, the use of fats as an energy source is discouraged. Fats can take the place of glucose, so if they are used as fuel the rate of glucose use will decline.

I hope it is clear that all these effects enable the body to make good use of glucose as and when it is available. The activities of many cells are simultaneously influenced by this hormone to achieve that purpose.

However, we still haven't answered the question of how insulin can do this from outside cells. The details of this are still being investigated, but it is thought that when hormones like this bind to their receptors, another substance is set free on the inside of the cell membrane and influences the cell's enzyme systems. This substance is called a **second messenger.**

I hope that this short account has provided you with a mental picture of the various ways in which the activities of any cell are controlled, both from within and without.

Let's revise some of these ideas through the next activity.

ACTIVITY 44

ALLOW 5 MINUTES

Let's return to the case study of the young man, Jason, with early-onset diabetes. Read the next instalment and then answer the questions which follow.

An Update on Jason

We last encountered Jason in a diabetic coma in Accident and Emergency. After appropriate treatment, the acidosis was corrected and his blood sugar brought down to a more normal level. He made a good recovery and has now been discharged from hospital. He is under the care of its Diabetic Clinic.

One of the tasks of the staff in this clinic is patient education. Jason needs to learn to recognise for himself the signs that his blood glucose is varying too much. The first thing they aim to do is to teach him about what will happen if his blood glucose falls too low – that is to say, if he becomes **hypoglycaemic.**

1 Under what circumstances would Jason become hypoglycaemic?

2 Can you give a biochemical explanation for hypoglycaemia, arising from the cause identified in the first question?

Commentary

There are several possible answers to these questions. They all involve an imbalance between carbohydrate intake and insulin availability.

For example, if Jason is inaccurate with his injection, he might give himself too much insulin. The biochemical effects of this would follow from insulin being absorbed on the receptors of the target cells, causing the 'glucose gates' to open and the second messenger to be produced. The cells would use up the available glucose quickly, and the level in the blood would fall. The brain would start to get short of glucose.

An alternative answer is that Jason might take his insulin, but neglect to eat the right amount of carbohydrate. In this case, the insulin would promote the uptake and use of glucose from his blood. If his diet did not replace the glucose, then his blood glucose level would fall and his brain would be deprived of glucose.

This is what happens when a person starts to 'go hypo'. It results in confusion and, sometimes, aggressive behaviour, as well as a range of recognisable physical symptoms such as sweating and hunger. It doesn't only happen in diabetes. Other people can experience low blood glucose, and its effects are something that health carers need to know about.

Jason's treatment and education will be aimed at getting a balance between the carbohydrate he eats and the insulin he injects, so that the level of glucose in his blood and in his cells remains as constant as possible. In other words, the treatment attempts to achieve homeostasis in a body which, because of the lack of a vital hormone, cannot do so unaided.

Summary

- Hormones are substances which are produced in one part of the body and have an effect elsewhere, on their target tissue or tissues.
- They exert this effect either by entering the cell and affecting the activity of the nucleus, or by remaining attached to the outside of the cell.
- The smaller hormones, such as steroids, enter the cell.
- Hormones with larger molecules, like insulin, remain outside the cell. They act by changing the rate of entry of raw materials into the cell, or through a second messenger, by affecting enzymic reactions inside the cell.
- The sum of all these effects is the maintenance of homeostasis and the efficient use of food raw materials.

Insulin is just one example of the many hormones which control cellular activity from outside. They too contribute to the maintenance of homeostasis. The end result of the activity of internal and external control is that all the enzyme systems in the cell work together in an integrated way.

Let's conclude this unit with a look at the integrated system of enzymic activity inside a cell.

4: Integrating all the cell's activities

I haven't yet emphasised that nearly all the reactions involving glucose which have been described in this unit are reversible. In other words, they can proceed in both directions. Glucose is split during glycolysis to form pyruvic acid, and pyruvic acid can be built up into glucose. These reversible reactions are shown by the double arrows in *Figure 20*.

Glucose is formed from several sources in the body, not just from starch and glycogen. It is produced from other carbohydrates in the diet   fructose, for example, the sugar found in fruit – and from some amino acids. Most of these reactions are reversible too, so that glucose is not only a source of fuel for the cell but it can be the raw material for synthesising many different compounds that the cell may need from time to time.

The first frame of *Figure 20* shows these relationships, and looks rather like a spider's web with glucose at its centre. We could, indeed, continue to add other sources and destinations for this key substance.

This is a very important diagram because it shows that there are many interconnected metabolic pathways which glucose can follow. All of them are subject to internal and external control which, in health, ensures that the constant, normal levels of the substances taking part in the pathways are maintained. If one of the pathways is disturbed, there will be an effect throughout the others.

However, glucose metabolism does not exist in isolation. Both amino acids (from protein), and fatty acids (from fat) are involved in metabolic pathways which link in to those of glucose. They too undergo reversible reactions through which fats and proteins can be built up into human tissues.

Let's look first at the links between protein and glucose metabolism.

Proteins from the diet are digested by amino acids and brought in to the cell. At the same time, the cellular proteins are continually wearing out and being digested by lysosomes, as we saw in Session Six. These proteins must be replaced.

The cell is usually well supplied with amino acids. These can either be used to rebuild cellular protein, or converted to glucose or one of the tricarboxylic acids. In both cases, the product can then take part in the metabolic pathways shown in *Figure 20*.

By adding the protein pathways to that diagram, we get to the second frame of *Figure 20*. From this, it can be seen that protein and carbohydrate metabolism are closely linked, and that a disturbance in one of the linked pathways will affect all the others.

Turning now to fat metabolism, we shall see that fats also share the pathways we have just described. Fats are digested to fatty acids which are, in turn, converted to the C_2 acid and enter the TCA cycle. Adding the pathways for fat to the others gives the third frame of *Figure 20*.

This final diagram shows how the breakdown and oxidation of all our food is linked to the synthesis of all the materials which make up our bodies. It represents the 'engine' which makes life possible.

The importance of these linked reactions for the health and well-being of our patients and clients cannot be overestimated. They constitute the fundamental processes keeping everyone alive. Ultimately, everything that goes into the body – as food or as treatment – influences this set of reactions. Every change in the internal environment affects it, and all the body's attempts to maintain homeostasis are directed towards keeping these reactions going. When a person is ill, this fundamental process may be threatened, but the body's survival mechanisms attempt to keep it going at any cost.

Knowledge of this fundamental set of reactions helps us to understand many of the effects of changes in diet or treatment.

For example, consider what happens to most of us just after Christmas, when we have enjoyed a more than adequate diet with plenty of protein and fat but, above all, an excess of carbohydrate in puddings, cakes and mince pies. Using the third

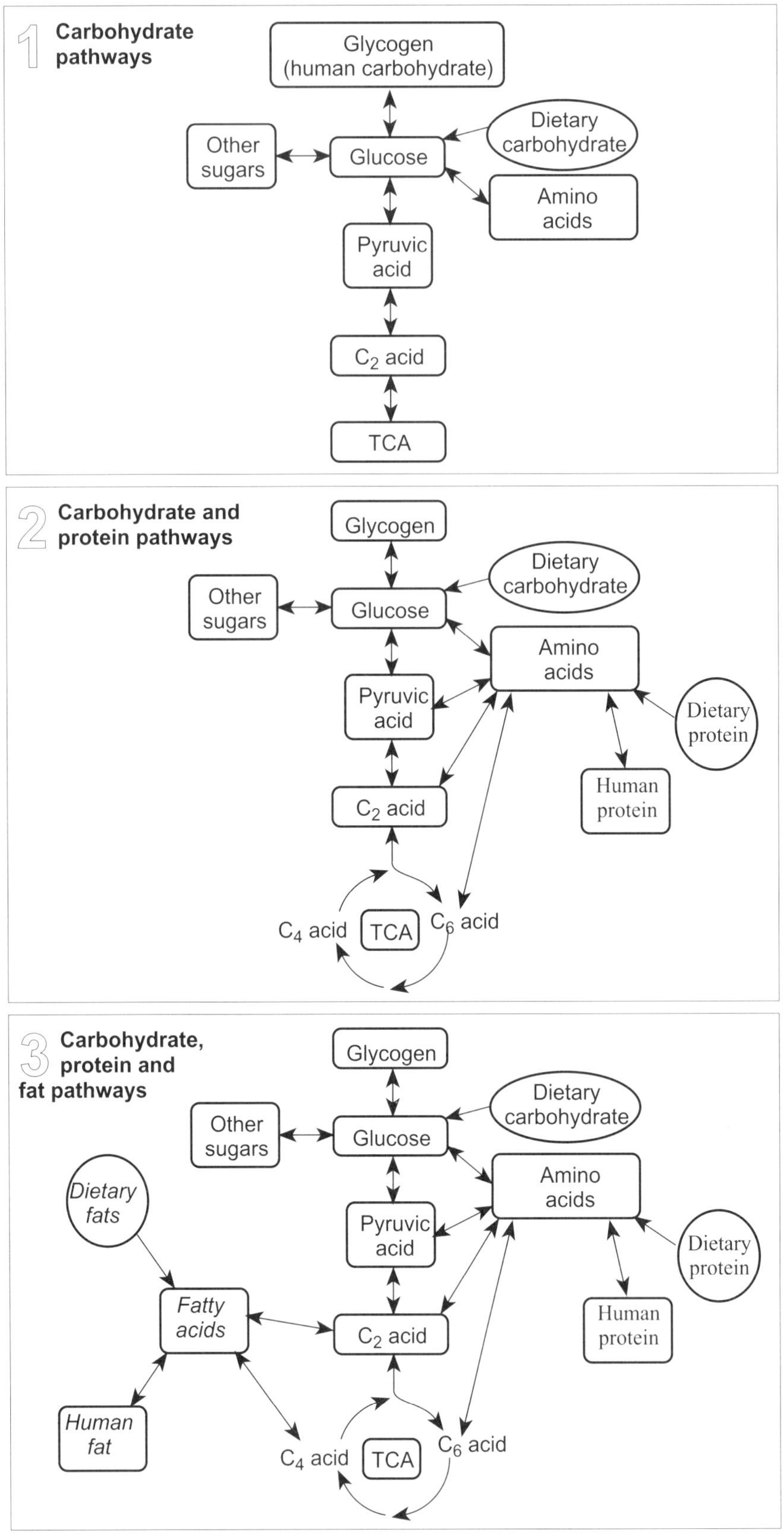

Figure 20 Integrated metabolic pathways

frame of *Figure 20*, we can work out that when we have enough protein, the glycogen storage space is full and we have produced enough ATP for our needs, there is only one place for the excess carbohydrate to go – to stored fat.

Final Self Assessment

Now that you have completed work on this unit, you may like to assess your progress and understanding.

- Complete the the Learning Review which appears at the end of the unit and compare your answers with those you gave before you started Session One.
- Reread *Resource 1* and look back at the notes and queries you made in *Activity 1* at the start of the unit to see how your level of knowledge has increased.

Learning Review

	Not at all	Partly	Quite well	Very well
Session One				
I can:				
● explain in general terms how different types of cell are adapted for their particular function.	☐	☐	☐	☐
Session Two				
I can:				
● identify the structures on which protein synthesis takes place in the cell	☐	☐	☐	☐
● describe the mechanism of protein synthesis	☐	☐	☐	☐
● explain how protein synthesis is controlled by the nucleus	☐	☐	☐	☐
● identify ways in which health care can support this function.	☐	☐	☐	☐
Session Three				
I can:				
● describe how some proteins function as enzymes	☐	☐	☐	☐
● identify what enzymes actually do	☐	☐	☐	☐
● describe the conditions that enzymes need	☐	☐	☐	☐
● explain how they can be affected by environmental and genetic factors.	☐	☐	☐	☐
Session Four				
I can:				
● describe how food materials are broken down into a relatively small number of basic materials	☐	☐	☐	☐
● explain how theses materials – especially glucose – get into the cell	☐	☐	☐	☐
● describe how this range of materials can be used to make all the cell's products.	☐	☐	☐	☐

	Not at all	Partly	Quite well	Very well
Session Five				
I can:				
• describe where the energy to drive cellular activity comes from	☐	☐	☐	☐
• explain what mitochondria are and how they function	☐	☐	☐	☐
• explain where mitochondria are thought to have originated.	☐	☐	☐	☐
Session Six				
I can:				
• describe how materials get in and out of the different parts of the cell	☐	☐	☐	☐
• explain how proteins are packaged for transport when they are complete	☐	☐	☐	☐
• outline how manufactured products get out of the cell.	☐	☐	☐	☐
Session Seven				
I can:				
• explain the need for control and integration within the cell	☐	☐	☐	☐
• describe methods of internal and external control of cell metabolism	☐	☐	☐	☐
• outline the role of hormones in external and internal control	☐	☐	☐	☐
• describe the action of insulin	☐	☐	☐	☐
• explain the concept of integration as applied to metabolic pathways in the cell.	☐	☐	☐	☐

Resources Section

Contents

RESOURCE I

Alan Dale, 'Diabetic Liaison', Community Outlook, May 1992

Diabetic liaison

Alan Dale explains how advice and reassurance from a diabetic liaison sister helped him deal effectively with a regime of insulin injections and blood glucose sampling after a diagnosis of diabetes.

It came almost as a relief to be to be told that I had a very high blood glucose level and that I was almost certainly suffering from uncontrolled diabetes. At least I now had an explanation for the severe mood swings I had been experiencing.

The nurse recommended that I use an electronic blood-testing kit based on an enzyme reaction.

The first symptoms appeared last April when I started to become very argumentative. I also developed oral thrush. I reasoned that the cause was the oral steroids I take for asthma. The fact that I have been taking steroids for eight years with no side-effects did not occur to me and, since the thrush responded well to Dakravin oral gel, I virtually ignored it. However, the infection reappeared after a few weeks and this time took rather longer to resolve. Again, I failed to recognise the significance.

I began increasingly to suffer from a nasty taste in my mouth, which is apparently associated with diabetes, although I have not found this mentioned in the popular textbooks. I found myself drinking more water than usual, but I thought this was because the weather was warm and because it got rid of the taste. I started to lose weight, but I was dieting, so this was to be expected.

Eventually I began hallucinating, initially in the form of vivid dreams, and then I became disoriented, especially at night. On one occasion I tried to walk down the stairs at home while perceiving them as a corridor. I did not hurt myself, but it was then that my wife called my GP. I confess I felt a certain relief that someone was going to help me.

I was admitted to the local hospital within two hours, where my blood glucose was found to be 45 mMoles per litre. The normal range is 4–7 mMoles per litre and the medical staff were quite surprised to find that I was still coherent, since such levels often result in a coma. On admission to the ward I was rehydrated using normal saline containing potassium ions. In addition, my blood glucose was monitored every two hours, at which time I also had insulin injections.

The next day I was visited by the diabetic liaison sister. My first degree is in biochemistry, but it is some years since I studied the subject, and it was useful to have someone explain the implications of diabetes.

I knew that before returning home I would have to learn to inject myself with insulin and, like most people, I was apprehensive about the procedure, so I arranged to do it the following morning and found that it was painless. In addition, I was shown how to take and interpret blood glucose samples so that I could monitor my own levels when I went home, which I did after three days in hospital.

The doctors put me on a regime of four injections of insulin a day, three rapid-acting ones and the last injection, just before bed, of a slow-acting type. On my discharge, the diabetic liaison sister suggested that I try injecting myself with a NovoPen. I found it easier to use and less conspicuous than the disposable syringes I had used in hospital. She also recommended that I rotate the site of injection to prevent the possibility of long-term tissue damage. In fact, the ability to inject into the subcutaneous fat of the stomach means that I can inject myself without attracting undue attention. This has proved useful on several occasions when I have taken insulin in crowded restaurants without people on adjoining tables noticing.

The sister explained to me that the advantage of the four-injections-a-day regime was that it is possible to delay mealtimes, at least by a reasonable time, in order to maintain a near as normal social life as possible. She also told me that the levels of insulin initially prescribed by doctors might need to be adjusted as the diabetes became more stable. The recommended method was to increase or reduce the levels by a maximum 2 i.u. per injection and to monitor the effect over a few days before making any further changes. This worked well and I now adjust the injection in accordance with the blood glucose level and the amount of carbohydrate that is in any meal.

I had been given a blood-testing kit based on a colour reaction, but I found that the results were rather difficult to interpret, since the ideal range of blood glucose, 4–7 mMoles a litre, was covered by one colour. This meant that it was difficult to decide whether the true level was near 7 mMoles a litre, in which case there was little danger of a hypo, or near 4 mMoles a litre, which indicated the need for food. The diabetic liaison nurse recommended that I replace this kit with an electronic version based on an enzyme reaction, which gives the results within 30 seconds and is accurate to the first decimal place. This has proved more satisfactory, and it is small enough to be carried in my jacket pocket.

I had my first, and so far only, hypo three days after leaving hospital. In retrospect, it was to be expected, since on that day I did not eat enough food and virtually missed a meal. The result was that I became unconscious for about 30 seconds just as I sat down for the evening meal. I found myself lying on the floor, with my dinner beside me on the carpet. My wife quickly got a glass of Lucozade and recovery was almost as rapid as the onset.

While in hospital I had read several books on diabetes which the diabetic liaison sister lent me. They all gave a list of the symptoms that should precede a hypo. Unfortunately, most of these symptoms did not occur in my case, possibly because I am on human insulin.

The symptoms I experienced just before the hypo, and which I have since come to associate with low blood glucose levels, are very subtle when compared with those described in the literature. They comprise mainly a tingling sensation in my fingertips, the tip of my nose and around my lips. The only other significant symptom is a slight mental confusion that seems to occur after the tingling has been present for about 10 minutes.

Once after buying something in a shop I realised that I could not remember how to put my credit card back into its case. When I took a blood glucose reading a few minutes later, I found the level to be 3.5 mMoles a litre, which is low. The level returned to normal within 20 minutes after eating a glucose tablet followed by a drink of orange juice and some sandwiches.

A mistake I made in the first few days after discharge from hospital was to assume that being hungry automatically meant that I was going to have a hypo. In fact, blood glucose readings taken when I was hungry were often in the normal range.

During the first week of convalescence, the diabetic liaison sister made a home visit. This was a great help, since various questions had occurred to me after I had left hospital and she was able to offer advice and reassurance. She also gave me her home telephone number. This proved useful when a few days later I was unsure whether I had injected myself correctly and she was able to tell me how to alter the next dose to compensate. I also telephoned her when going out for the evening meant that I would be eating several hours later than usual. Her advice was very useful and meant that my blood glucose levels stayed in the normal range. I was told to have a small snack, perhaps two biscuits or a sandwich, at the normal mealtime and to take insulin before the actual meal, whatever time that was. I then took the normal dose of slow-acting insulin before bed. Since then I have managed to resolve the various problems myself, but it is reassuring to know that I can always contact her if I do have any difficulties.

On my return to work, I had to convince my colleagues that I wasn't going to have a hypo at any moment. Indeed, a member of the first-aid team asked where I kept my insulin, so that he could inject me if I fainted. I explained to him that this was an inappropriate and potentially life-threatening action.

Obviously I need to take reasonable precautions against having a hypo, especially when driving. I always carry several digestive biscuits and glucose tablets. I also measure my blood glucose level just before leaving work to ensure that there is no danger of becoming too low before I reach home. As an additional precaution, I always have a packet of biscuits in the car in case I get badly delayed by traffic or the car breaks down.

I informed the Vehicle Licensing Authority at Swansea of my diabetes, since not to do so renders one's driving licence invalid. The authorities were sympathetic and, as is common practice, changed my licence to one valid for three years, the renewal being granted only with medical approval. It was also essential to inform my car insurance company, since diabetes constitutes an increased risk. The company did not increase the premium, but some have been known to put very high loadings on diabetic drivers, so it is useful to check before renewing a policy.

My wife and I have just returned from a holiday in New Zealand, during which time

Essential Advice for Travellers

- Take enough insulin and test strips for the whole trip, with a small surplus for accidents.
- Pack the insulin and test strips in two separate items of hand luggage, with one set carried by your partner if you are travelling together.
- Check that your insulin and test strips are available in the country you are visiting (the British Diabetic Association is very helpful in providing this information).
- When travelling long distances by plane, ignore the normal injection regime and inject a small dose before each meal. Revert to the normal regime on arrival. Blood-testing before each airline meal is a great help.

I experienced no difficulties in maintaining good blood glucose control. When travelling abroad, especially over such a distance, the problem of maintaining one's insulin regime arises, particularly when crossing time zones. The diabetic liaison nurse suggested that I ignore the usual timings and inject a slightly reduced dose of rapid-acting insulin before each meal. This worked very well, and my blood glucose levels showed little deviation from the normal values.

Before travelling, I checked that the insulin and test strips I use were available in New Zealand. One other precaution I took when travelling was to ensure that a spare NovoPen and spare cartridges of both insulins were packed in a different piece of hand luggage to my normal supplies, just in case the airline managed to lose something. I took enough insulin and test strips for the holiday with some extra in case of accidents.

The only incident throughout the holiday occurred when I allowed myself two extra mince pies (well, it was Christmas), and found my blood glucose level had risen above 10 mMoles a litre. That proved a salutory lesson not to become over-confident about my diet.

I have six-monthly check-ups, when my retinas are checked, although little emphasis is given to foot care, perhaps because of my age (46).

It is tempting to worry about the potential long-term effects of having diabetes, even though these are better understood these days and most can be treated. But after the initial shock I resolved that the most important thing was to attain the best possible control and to lead a normal life. In this it is very helpful to have the back-up of the diabetic nurse specialist. I have only had to phone her twice so far, but it is reassuring to know that she is on hand so that I can get immediate practical advice.

The genetic code

RESOURCE 2

Lyne, Dr P, Genetics, Open Learning Foundation, 1994

What is DNA?

The variation which is present at birth, which we can identify as **genetic variation**, must have its origin in the fertilised egg at conception. This is the starting point from which every person grows. Therefore, it must contain all the information needed to make that person into a unique individual. Let us first consider the nature of that information, in order to understand how it functions.

The fertilised egg, like most other cells in the human body, contains a **nucleus**. The nucleus is a distinct region of the cell, separated from the cell contents by a membrane and acting as a control centre for the cell's activities.

Viewed through a microscope, the **nuclei** of living cells look spherical and move around inside the cells as the contents of the cells move.

Every nucleus contains substances called **nucleic acids**. These are so-called because they were first discovered in nuclei and it was thought that they did not occur anywhere else. This has since been found to be incorrect, but the name remains both as an accident of history and because they are more abundant in nuclei than in other parts of the cell.

There are two types of nucleic acid which, although constructed from similar basic materials, have different functions in the cell. Their correct chemical names are **deoxyribonucleic acid** and **ribonucleic acid**, but they are usually referred to by the abbreviations **DNA** and **RNA**.

You may well have come across DNA in your reading, or have heard it reported in the media in connection with some current controversy.

ACTIVITY 7

At the moment, when you hear the term DNA, what comes to mind? Make a note in the space below of the things you associate with DNA, and jot down anything which you find puzzling about it.

- Thoughts associated with DNA:

- Puzzling features:

Commentary

Many people who answer this question come up with comments such as:

'It's something to do with genetics'
'It enables cells to replicate'
'It's to do with genetic engineering, whatever that may be'
'I know it's a double helix discovered by an American, because I saw a TV play, but I've no idea what that means'
'I've heard words like cloning, fingerprinting and pre-natal diagnosis used along with it'
'It can be used to identify a child's father, can't it?'.

Ideas of this kind arise from everyday experience – from the newspapers, the television and in general conversation. However, after completing this section, you will have an increased understanding of the nature of DNA and of its significance in many aspects of our lives.

DNA and the genetic code

As you made your own list, I wonder whether you added one item which is missing from the one given above. Did you make the point that *DNA carries genetic information*?

I have already introduced this idea, and will now explain it in more detail. Since this is not a course for scientists, I am not assuming that you have any understanding of chemical formulae and structures. I will therefore be using metaphors to illustrate some of the ideas that you will need to think through now.

When describing how DNA works, a useful metaphor is that of a book of instructions. We can visualise a person's DNA as an instruction manual for constructing and operating that particular person.

Like any other book, a manual is composed of sentences and words written in a language – Welsh, Urdu, English and so on – which has meaning to anyone who understands the rules of that language. So a language is a form of code. In the kind of language that you and I understand, the meaning is conveyed by letters of the alphabet, their sequence and the way that they are grouped together.

The DNA manual is also 'written' in a language or code – the genetic code. It exists in the actual structure of DNA, which we could think of as a 'living book'. The 'letters' in which it is written are the sub-units from which the DNA is built up and the meaning is conveyed by the sequence of these sub-units and the way that they are grouped together.

The English alphabet contains 26 letters. Other languages have more or less, but always an appreciable number. There are only four different sub-units in DNA and so the DNA language has only four 'letters'. All DNA is composed of long chains of these sub-units. They are usually identified by the initial letters of their chemical names – A, T, C and G. So a section of DNA chain could be shown as a series of these four letters, indicating the order of the sub-units in that part of the chain. For example ... ATTGGCTAGT ... and so on.

This series of letters spells out a section of the genetic code. Although we ourselves cannot read it, it has meaning inside the living cell. In order to make this meaning clear, I am going to draw on some material from the biochemistry unit.

All living cells carry out complex activities which make up the actual process of living. These activities are brought about by **proteins** – which can be thought of as the cell's machinery. The kind of work that any particular cell can do depends on the proteins which it produces at any given time. So it should be clear that proteins are the key to the cell's activities, and ultimately to the well-being of the whole body. Thus the genetic information carried in the nucleus has to control this most fundamental of processes, the manufacture of proteins.

This is exactly what the DNA code does. All proteins are built up from small building blocks called **amino acids**. The nature of a protein depends on the order in which the amino acids are joined together, and the DNA code spells out the order in which this is to happen when a protein is made.

Just as the letters in an instruction book are grouped together to form words, so the DNA sub-units are grouped to form units of meaning which we can think of as 'words'. Unlike the book, all the DNA words are the same length. They are all three letters – or three sub-units – long. For that reason, they are termed **triplets** and the DNA code is known as a **triplet code**.

Figure 5 shows the relationship between the DNA words and the proteins which are made in the cell.

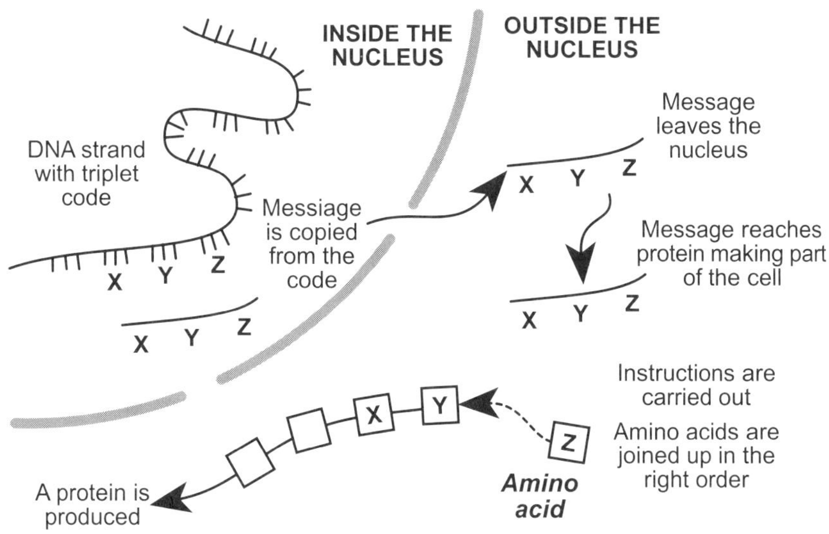

Figure 5 The relationship between DNA and protein manufacture

Inside the nucleus the DNA message is copied onto **RNA** - the second type of nucleic acid. RNA is very similar to DNA , but it exists in smaller fragments which can easily pass out of the nucleus (through pores in the nuclear membrane) and so transmit the message written in the DNA. The way in which the code is copied is described in section 4.1 and the synthesis of proteins is detailed in the biochemistry unit of this series.

The parts of the cell which manufacture proteins recognise triplet words. Each word specifies where a particular amino acid should go when the cellular machinery is making a protein. Each one means:

Put amino acid X here when you make a protein

The piece of code shown in *Figure 5* contains three words, reading from left to right:
att/ggc/tag/t, and so on. Each word specifies a different amino acid.
So suppose that:

ATT	=	amino acid 'X'
GGC	=	amino acid 'Y'
TAG	=	amino acid 'Z

then this piece of code will instruct the cell to make a protein in which x, y and z are joined together in that order.

In the next activity, I want you to write a piece of genetic code for yourself, and begin to think about the effect of making a mistake.

ACTIVITY 8

The DNA code word for amino acid B is ATA.

1 Write a piece of code which will instruct the cell to join three Bs together.

2 Then write it out again, but this time make a deliberate mistake with the fourth letter along from the left. You could miss it out, put it in twice, or change it for another letter.

3 Now look at your piece of code and see what the effect will be on the instructions which it gives to the cell

Commentary

The piece of code you're looking for is:

... ATA ATA ATA ...

Your answer to the first part of the activity depends on the principle that the genetic code is based on groups of three letters, always read in the same direction, which I have shown as left to right.

This principle is not always easy to follow the first time, because it requires us to deal with abstractions. So if you didn't get the answer right this time, don't worry – you're certainly not alone! Try reading through 'The genetic code' again, and follow the argument step by step.

Here are some of the things you might have done for the second part of the activity:

- delete a letter ATA TA ATA
- add a letter ATAA ATA ATA
- replace a letter ATA GTA ATA

In each of these cases, you will find that – reading from left to right in groups of three – you get a different set of 'words'.

Deleting a letter gives ATA TAA TA?

This gives one amino acid B, followed by two different amino acids.

Addition gives you ATA AAT AAT A??

Again, the first amino acid in the protein chain is amino acid B, but the others will be different.

Replacing a letter gives ATA GTA ATA

The first and third amino acids are B, but the middle one is different.

The result of all three kinds of mistake is a piece of code which is different from the original sequence. If this happens to the code for part of a protein, the resulting protein will obviously be slightly different from the original.

Variation in the code

What you have just done is a pencil and paper exercise which mirrors what can and does happen in the cell. It shows how a change in the code changes the order of the amino acids from which the proteins are made up.

But does this matter?

In the next activity, we'll look at an example of the effects of a change in the code, and the resultant change in some very important proteins.

ACTIVITY 9

Please read through the following case study, and then note your answers to the questions that follow it.

Karima is a very ill eight-month-old baby. She was born in London to parents who arrived there from Pakistan three years ago.

She was a healthy baby, delivered normally, and was well for the first three months of her life. She began to show a failure to thrive, and has now been admitted to a children's ward with profound anaemia. She has recurrent fevers and an enlarged spleen, causing her abdomen to be very distended.

Her parents are very worried because Karima had a little brother who died at the age of three, after showing similar symptoms a few months after he was born. As time went on, he failed to grow and his legs became deformed.

Karima's parents have two other children, both well. Her mother has just found that she is pregnant again, and both parents had younger sisters who died in infancy of a severe type of anaemia.

Karima is diagnosed as having a type of blood disorder called thalassaemia, and is given immediate treatment for the anaemia. Her mother is offered a test to check the DNA of the new baby before it is born.

1 Now that you have read about Karima's illness – thalassaemia – identify the systems of the body which it affects.

2 Do you think that this illness is caused by genetic or environmental factors, or by a combination of the two?

Commentary

Your answers should have been along the following lines.

1 The illness affected Karima's blood system, giving rise to the anaemia and the enlarged spleen. In her brother's case, it affected his skeleton too, since his limbs were deformed. As her brother died, there must also have been far-reaching effects on other body systems.

2 At first sight, it seems as if this condition is due to the genetic factors. It runs in this family, and the mother is offered a genetic test on the unborn baby. However, the fact that the illness started several months after birth is a bit puzzling. Is this really a genetic effect or could environmental factors have been involved? It is difficult to say without further information.

A fuller explanation of Karima's case

Thalassaemia is an inherited illness. It results in a failure to make the oxygen carrying part of the blood, **haemoglobin (Hb)**, properly. This results in anaemia.

The unborn child has a type of haemoglobin (**foetal Hb**) which is slightly different from the adult type (**adult Hb**). It picks up oxygen more easily, and so can extract oxygen from the mother's blood as it flows through the placenta. When the baby is born, it continues to use foetal Hb for a few months, but soon needs to start making adult Hb. This is when the trouble starts for a child with thalassaemia.

The problem is that part of the genetic code responsible for the production of adult Hb contains a mistake in people who suffer from this illness. The part responsible for foetal Hb is not affected. The baby is well at first, but cannot make adult Hb when the time comes. The child therefore becomes more and more anaemic.

In an attempt to increase the output of this vital protein, the blood-forming tissues of the body increase in volume. Thus the spleen and bone marrow (the place where red-blood cells are made) get larger, causing the abdomen to swell and bones to become distorted. Without treatment, this form of the disease leads to severe illness and death.

As you read the description of the severity of these symptoms, you may have wondered about the nature of the genetic 'mistake' which could have such far-reaching effects. This is known to be due to a single mistake, like the one you made in *Activity 8*. It is actually due to a replacement in the code, which changes a single amino acid in part of the adult Hb molecule. Because this oxygen-carrying substance is so central to all the activities of living, this apparently insignificant change has very widespread consequences.

I provided this example to demonstrate what a small change in a crucial part of the genetic code can do. It produces a malfunction, and therefore constitutes a mistake in the code.

However, there are many ways in which the code can vary without producing adverse effects – just slight differences of the kind that we have been observing in healthy people.

Within each nucleus there is a great deal of DNA. It is calculated that, if all the DNA in a nucleus were to be unravelled and stretched out, it would be approximately two metres in length and would contain something like three billion (3000,000,000) sub-units.

There are so many possible places for the code to vary that each person has a unique genetic code – their **DNA fingerprint**. No two people, apart from identical twins, have exactly the same sequence of units in their DNA. Nowadays, it is possible for geneticists to work out the sequence in the nuclei of living or dead cells. This is why small samples of hair or skin can be used to track down criminals. The sequence of the DNA code found in their cells can be matched against those of people suspected of the crime.

Here we have an explanation of the source of the genetic diversity of the human race. Each person starts life with a unique set of genetic code in the fertilised egg. This leads us to ask more questions, such as how do the differences in the code arise? We'll consider this subject in the next section, but first here is a summary of the main points covered in this section.

Summary

- The nuclei of all cells contain DNA.
- DNA carries genetic information in the form of the genetic code.
- This information determines the manufacture of proteins.
- The code is written in the sequence of DNA sub-units (A, T, G and C).
- Code 'words' are groups of three sub-units called triplets.
- Each triplet word specifies the position of an amino acid in a protein.
- Small changes in the code are responsible for variation.
- Some changes cause malfunctions.
- Each person has a unique genetic sequence.

RESOURCE 3

Stephen Thomas, 'Dealing with fungating lesions', Community Outlook, May 1992

Dealing with fungating lesions

Little practical advice is available about the management of these uncommon but distressing wounds. Stephen Thomas describes some research findings

Fungating wounds, although relatively uncommon, present major problems both to patients and those responsible for their care. They are most often associated with carcinoma of the female breast but can also arise from other internal malignancies and may develop at the site of the original lesion or elsewhere as a result of infiltration of skin epithelium by cancerous cells. From these a tumour develops and increases in size until it loses its vascularity when capillary rupture occurs, followed by necrosis.[1,2]

Radiotherapy may reduce the size of the resultant area, but such wounds are frequently not expected to heal and may bleed profusely when dressings are changed. The wounds often become infected with anaerobic organisms that produce volatile fatty acids with a pungent odour that can be a source of great embarrassment and distress to the patient and his or her relatives.

A review of the literature reveals little practical information or advice on the management of fungating wounds, as most of the papers that have been published have tended to concentrate primarily on methods of controlling the smell produced by such wounds.

In 1976 Butcher and Maggs described the odour-absorbing properties of a charcoal cloth developed by the Chemical Defence Establishment at Porton Down.[3] This, together with cotton gauze and a layer of water-repellent fabric, was formed into absorbent pads that were found to be very effective in containing the odour from fungating wounds. This simple product proved to be the forerunner of a number of 'deodorising' dressings, the construction and use of which have been reviewed previously.[4]

Numerous other methods of preventing or eliminating odour have also been described, and these include the use of icing sugar and honey.[5,6] It is believed that the latter act by producing a hyperosmotic environment within the wound that inhibits bacterial growth, thus preventing odour formation.[7] Live yoghurt is also sometimes applied in an attempt to encourage overgrowth of pathogenic organisms by lactic acid bacteria such as Lactobacillus bulgaricus and Streptococcus thermophilus.

In 1980, Ashford et al described the successful use of oral metronidazole in a patient with an offensive ulcerating mammary tumour.[8] The value of this treatment was soon confirmed by others.[9] Similar success in the management of malodorous gynaecological tumours was reported by Dankert et al[10] who demonstrated that metabolites such as propionic, isobutyric and isovaleric acids (produced by anaerobic organisms) were eliminated from the vaginal discharge of affected patients treated with oral metronidazole.

These early reports were subsequently supported by the findings of a double-blind, randomised, crossover trial which confirmed that, compared with a placebo, oral metronidazole was highly effective at reducing odour and eliminating anaerobes or anaerobic metabolites from the wounds of breast cancer patients.[11]

Unfortunately, the long-term use of oral metronidazole may cause neuropathy, and patients who are receiving systematic therapy must refrain from consuming alcohol. Some researchers have therefore investigated the use of topical formulations. Sparrow et al refer to the use of saline soaks containing the drug applied to the wound twice daily,[9] and a similar technique was reported by Gomolin and Brand[12] who used an aqueous solution of metronidazole in the treatment of necrotic pressure sores, following the work of Jones et al who used the same solution of the treatment of pressure sores, diabetic ulcers and abscesses.[13]

A more convenient topical formulation was described by Allwood, which consists of an aqueous hypromellose gel containing metronidazole 0.8%.[14] When tested on 68 fungating wounds, the gel was found to eliminate odour from 34 of them and have a reasonable effect on a further 31.[15] Only three wounds treated with the gel failed to improve. This formulation, called Metrotop, has recently obtained a product licence and is available on prescription.

Cotton gauze may produce fresh damage when removed from the surface of a wound, causing bleeding.

An alternative formulation consisting of metronidazole 0.8% in Scherisorb intrasite gel is also used,[16] although this is not commercially available and therefore has to be prepared extemporaneously. Laboratory studies on both gel formulations suggest that, despite the fact that metronidazole is traditionally associated with the treatment of anaerobic infections, in the concentrations used topically it may also have an effect upon a range of aerobic organisms, some of which can cause wound infections or lead to odour.[17]

WOUND TYPE	**RECOMMENDED DRESSINGS**
Bleeding wounds	Alginate sheet dressings (Kaltostat or Sorbsan).
Shallow, lightly exuding wounds	Hydrocolloid sheets, semipermeable films, Lyofoam, Sorbsan SA or Kaltoclude. Alternatively, Silicone NA or Tegapore used with a secondary dressing.
Heavily exuding wounds	Sorbsan, Kaltostat or Kaltostat Fortex covered with an absorbent pad. Alternatively, Sorbsan Plus or Allevyn sheet dressing may be applied.
Cavity wounds	Alginate rope or packing. Allevyn cavity wound dressing or Silastic foam.
Malodorous wounds	Dressings containing activated charcoal, such as Actisorb Plus, Kaltocarb, Lyofoam C, topical metronidazole (Metrotop), intrasite gel (Scherisorb) containing metronidazole or sugar paste.
Infected wounds	Topical metronidazole (Metrotop) or intrasite gel (Scherisorb) with metronidazole, sugar paste, alginate dressings, products containing iodine or povidone iodine, such as Inadine or Iodosorb.
Necrotic/sloughy wounds	Intrasite gel (Scherisorb), Sorbsan, sugar paste, Debrisan paste, Iodosorb.

Table 1 Dressings for the management of fungating wounds

Some of the practical problems of managing fungating wounds were discussed by Bale and Harding, who described the advantages of a foam elastomer dressing (Silastic foam) for this application, following their experience in a specialist wound clinic.[18]

An interesting literature review, covering many aspects of the care of cancer patients who develop fungating and/or ulcerating lesions (including psychological factors), was published by Ivetic and Lyne in 1990.[19]

In addition to the above, a few anecdotal case studies and reports of different treatments are also available,[2,20,21] but reports of controlled clinical trials of different treatments are hard to find.

In the absence of this information, a survey was undertaken to obtain the opinion of medical and nursing staff on the value of dressing treatments in current use. Questionnaires were sent to all radiotherapy departments in the UK, as well as other specialist centres and oncology units, asking respondents to indicate the value of each of some 50 dressings using a simple analogue scale. A total of 328 forms were sent out, followed by a reminder letter as appropriate, and 117 (35.7%) were returned, three of which were unusable owing to incorrect completion or the provision of inadequate information.

Full details of the results of the survey have been published elsewhere,[22] but the principal conclusions are summarised here. Some simple guidelines for the management of different types of fungating wounds have been produced from these results, which may be of value to nursing staff required to treat fungating wounds on an infrequent or irregular basis.

Although numerous solutions are available for cleansing wounds, the survey revealed a surprisingly low usage of antiseptics. Most people (54%) relied upon a solution of sodium chloride 0.9% (normal saline), and only 7% applied either Savlon or hypochlorite solutions.

Some fungating wounds bleed easily on contact. In these situations, dressings made from alginate fibre (Kaltostat or Sorbsan) were used by 56.7% of respondents. Cotton gauze was used by 18.7%, but although this may help to produce haemostasis, unlike the alginates which can be removed without causing further trauma, cotton gauze may produce fresh damage when removed from the surface of a wound, causing secondary bleeding.

The dressings that scored most highly in the survey for the management of low-exudate wounds included the simple, low-adherent absorbent dressing pads such as Melolin, Telfa and Lyofoam. Granuflex (where the size and shape of the area permit) and film dressings such as Bioclusive and Tegaderm were also recommended. Traditional wound contact materials such as paraffin gauze, both plain and medicated, also achieved reasonable scores.

Absorbent pads and Gamgee tissue gained the highest scores for heavily exuding wounds, but Sorbsan, Sorbsan Plus and Kaltostat also scored highly.

Dressings containing activated charcoal, such as Actisorb Plus, Kaltocarb, Carbonet and Lyofoam C, were widely used for malodorous wounds, but topical metronidazole was also highly recommended. The survey revealed that Sorbsan was considered to be very useful in the treatment of malodorous wounds, although the reason for this is not obvious.

In the management of infected wounds, metronidazole was highly recommended by many respondents in the survey.

Dressings used in the management of necrotic/sloughy wounds included the polysaccharide bead products such as Debrisan paste, together with intrasite Scherisorb gel, Granuflex granules and Sorbsan.

With regard to the management of infected wounds, the survey produced some interesting, if somewhat unexpected, results. Metronidazole, although not on the list of dressings included in the original questionnaire, was highly recommended by many of those who took part, even though only a small proportion of respondents actually gave the item a score.

The most popular dressing treatment, however, was Sorbsan, which was rated higher than either the medicated tulles or dressings containing iodine or povidone iodine, products which are traditionally associated with the treatment of infected wounds.

Examination of the overall scores achieved by each dressing clearly indicates that the most popular products for the treatment of all types of fungating wounds are the alginate dressings (particularly Sorbsan), followed by Lyofoam and Actisorb Plus.

Table 1 contains recommendations for the management of fungating wounds, but it should be emphasised that these are not based upon the results of controlled clinical trials with the materials concerned, as this information is simply not available. They are drawn principally from the results of the survey. In some instances they also reflect the personal views of the author on the value of certain new dressings or treatments in the management of the different wound types and are based upon a knowledge of the properties and performance characteristics of the materials concerned. It is also not intended that these recommendations should be followed blindly but, rather, be used as a guide during the dressing selection process.

References

1 Foltz, A.T. Nursing care of ulcerating metastatic lesions. Oncology Nursing Forum 1980; 7: 8–13.

2 Sims, R., Fitzgerald, V. Community Nursing Management of Patients with Ulcerating/Fungating Malignant Breast Disease. London: Royal College of Nursing. 1987.

3 Butcher, G., Maggs, F.A.P. The treatment of malodorous wounds. Nursing Mirror 1976; 142: 6.

4 Thomas, S. Wound Management and Dressings. London: Pharmaceutical Press, 1990.

5 Thomlinson, R.H. Kitchen remedy for necrotic malignant breast ulcers. Lancet 1980; 2: 594.

6 Keast-Butler, J. Honey for necrotic malignant breast ulcers. Lancet 1980; 2: 809.

7 Chirife, J. In-vitro study of bacterial growth inhibition in concentrated sugar solutions: microbiological basis for the use of sugar in treating infected wounds. Antimicrobial Agents and Chemotherapy 1983; 23: 766–773.

8 Ashford, R.F., Plant, G.T., Maher, J. et al. Metronidazole in smelly tumours. Lancet 1980; 1: 874–875.

9 Sparrow, G., Minton, Rubens, R.D. et al. Metronidazole in smelly tumours. Lancet 1980; 1: 1185.

10 Dankert, J., Holloway, Y., Bouma, J. et al. Metronidazole in smelly gynaecological tumours. Lancet 1981; 2: 1295.

11 Ashford, R., Plant, G., Maher, J. et al. Double-blind trial of metronidazole in malodorous ulcerating tumours. Lancet 1984; 1: 874–875.

12 Gomolin, I.H., Brandt, J.L. Topical metronidazole therapy for pressure sores for geriatric patients. Journal of the American Geriatric Society 1983; 31: 710–712.

13 Jones, P.H., Willis, P.H., Ferguson, I.R. Treatment of anaerobically infected

pressure sores with topical metronidazole. Lancet 1978; 1: 214.

14 Allwood, M., Hutchinson, D.I. Commercial version of metronidazole gel. Pharmaceutical Journal 1987; 238: 504.

15 Newman, V., Allwood, M., Oakes, R.A. The use of metronidazole gel to control the smell of malodorous lesions. Palliative Medicine 1989; 3: 303–305.

16 Thomas, S. Treating malodorous wounds. Community Outlook 1989, October, 27–30.

17 Thomas, S., Hay, N.P. The antimicrobial properties of two metronidazole medicated dressings used to treat malodorous wounds. Pharmaceutical Journal 1991; 246: 264–266.

18 Bale, S., Harding, K. Fungating breast wounds. Journal of District Nursing 1987; 5: 4–5.

19 Ivetic, O., Lyne, P.A. Fungating and ulcerating malignant lesions: a review of the literature. Journal of Advanced Nursing 1990; 15: 83–88.

20 Johnson, A. Dressings and Rehabilitation: A Case Report Illustrating the Benefit of a New 'Odour-Eating' Dressing. Sittingbourne: Ultra Laboratories, 1988.

21 Neal, K. Treating fungating lesions. Nursing Times 1991; 87: 36, 84–86.

22 Thomas, S. Current Practices in the Management of Fungating Lesions and Radiation Damaged Skin. Bridgend: Surgical Materials Testing Laboratory, 1991.

The information in this article has been taken from a booklet by Stephen Thomas, *Current Practices in the Management of Fungating Lesions and Radiation Damaged Skin*, which is available from the Surgical Materials Testing Laboratory, Bridgend General Hospital, Quarella Road, Bridgend, Mid Glamorgan CF31 1JP. The price is £5 including postage.

Stephen Thomas, BPharm, PhD, MRPharmS, is director, Surgical Materials Testing Laboratory, Bridgend General Hospital, Mid Glamorgan.

Debriding agents

RESOURCE 4A

Extract from article by Anne Lister in Wound Management, Vol 1: 3

Item 1

Our most highly recommended cleansing agent is normal saline 0.9% w/v. It is damaging to neither wound nor surrounding skin.

For sloughy wounds or wounds covered with hard necrotic tissue, the application of *Varidase* may be useful as it lifts the necrotic tissue. When reconstituting *Varidase*, care should be taken not to shake the vial as this may render the enzymes (Streptokinase and Streptodornase) impotent. For the dry necrotic wound, reconstitute *Varidase* with 20 ml normal saline 0.9% w/v and inject the solution directly under the eschar. For the deeper sloughy wound, dissolve Varidase in 5ml water, mix with 15ml KY gel and apply direct to the wound.

RESOURCE 4B

Extract from article on wound care by Arthur Johnson in Nursing Times

Item 2

If the necrotic material is particularly dehydrated and hard, debridement can be speeded up by employing the enzymes streptokinase and streptodornase under occlusion. Gently mix an ampoule or vial of *Varidase* powder with 5–10ml KY jelly, and apply directly onto the necrotic material.

RESOURCE 4C

Extract from 'Varidase' advertisement

Item 3

Indications: Cleansing of necrotic and infected wounds, suppurative surface lesions such as ulcers, pressure sores, burns. Dissolution of clots in the bladder or urinary catheters.

- Dosage and administration: Treatment with the contents of one vial repeated once or twice a day. See data sheet or package leaflet for full directions for use.
- Contra-indications: Active haemorrhage.
- Precautions: For local use only. Do not use intramuscularly or intravenously.
- Side effects: Infrequently, allergic reactions may occur.
- Presentation: Vials of dry, sterile powder, which requires storage under refrigeration (2–8°C), each containing 100,000 units of Streptokinase and 25,000 units of Streptodornase.

RESOURCE 5

Ray Newton, 'Testing, testing', Community Outlook, January 1987, pp 16–18

Diabetes

Testing, testing

Good blood glucose control is the key to a healthy life for diabetics, but some find urine testing messy and imprecise, according to Ray Newton. He explains why blood testing is more accurate, and how patients can be encouraged to do their own testing in the community, with back-up from nurses and the latest equipment.

A successful life as a diabetic depends on achieving good blood glucose control and there is increasing evidence that this is one important way of preventing long-term complications. Education has been directed at encouraging people with diabetes to use the results of urine tests and more recently home blood tests, to assess their diabetic control and make necessary adjustments to treatment. However, success depends not only on the know-how but also the will to do regular tests.

It is crucial that diabetics are made aware that tests results are primarily for their own use and are not purely to satisfy nursing or medical staff. Many diabetics take the attitude 'it's all right for them they don't have to do tests'. Unfortunately the record of tests often becomes a point of confrontation at the clinic and we are all aware of falsified results. Several reports confirm these compliance difficulties.[1,2] Discussion of test recording is obviously an integral part of the diabetic clinic consultation but confidence and trust between doctor and patient will come only from a common sense approach to home testing.

People are more likely to do a test which is simple and quick – one that can be done at half time in a football match, during an exam or in the factory on night shift. Blood tests can be performed in all these circumstances and urine tests can be done in some of them, where chemical test strips such as Diastix or Diabur 5000 are used. Many diabetics find urine testing messy and embarrassing, and these tests are only an indirect indication of blood glucose level. Variability of the renal threshold (the blood level at which glucose spills into urine) for example in pregnancy or in the elderly, and the fact that a negative urine test may correspond to a blood glucose value anywhere between 0–10 mmol/l, inevitably suggests that urine testing is imprecise.

Clearly, blood testing is more accurate and many diabetics now prefer it. It is impor-

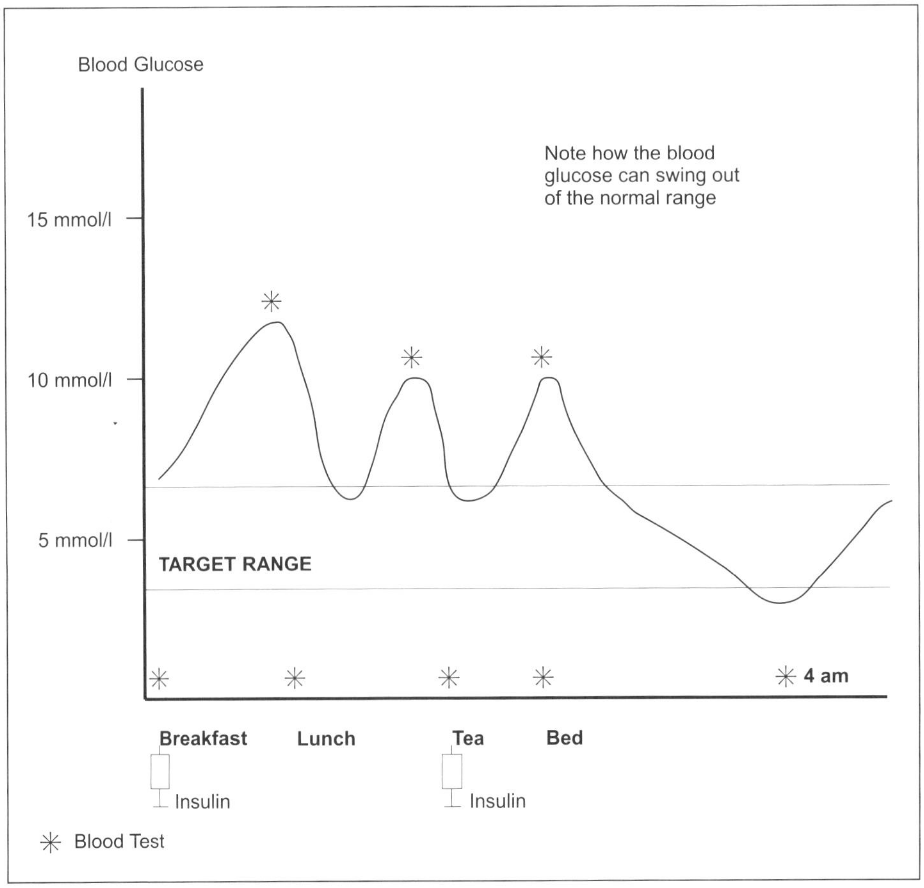

Table 1 Useful times to perform blood tests

	Type	Special points	Approx cost		Obtainable from
Lancets	Unilet	Fits most devices	200	£6.00	Owen Munford
	Monolet	Fits most devices	200	£6.00	Mariner Medical
	B-D Microfine	Fits Autolance claim of greater comfort			Hypoguard
Pricking devices	Monojector	Pencil Type (Lancet hidden)		£6.00	Mariner Medical
	Autolet	Comes in form with timer		£6.00	Owen Munford
	Autoclix	Adjustable platform		£8.00	Hypoguard
	Haemalet	Pencil type		£6.00	Hypoguard
	Autolance	Pencil type		£6.00	Hypoguard
Strips	Visidex II		50	£10.00	Hypoguard, Ames
	Bm Test Glycemie 1-44	Can be split for economy	50	£14.00	Hypoguard, Boehringer, Owen Munford
	Glucostix				Ames, Hypoguard, Owen Munford
Meters	Refloflux II	Only BM strips Reads wide range		£100.00	Boehringer, Owen Munford
	Glucometer	Only Glucostix Memory system available Fast test		£89.00	Ames, Owen Munford
	Glucocheck™SC			£80.00	Medistron, Owen Munford
	Hypocount MX	Smallest available Memory system Available for all strips		£85.00	Medistron, Owen Munford Hypoguard

Table 2 Blood glucose monitoring equipment

tant to accept, however, that home blood glucose monitoring is not suitable for everybody:

- Some find the discomfort of finger pricking to obtain blood unacceptable.
- Blood testing strips are expensive if clinics are unable to provide them free of charge (the government may allow prescribing on the Drug Tariff in the near future).
- Insulin dependent diabetics who test blood, soon realise that their glucose level may swing considerably throughout the day and may become alarmed by occasional high readings unless these are carefully explained. Non-diabetics maintain their blood glucose within quite strict limits of between 3.5 mmol/l and 6.5 mmol/l and this is the usual target range for pre-meal tests.

When and how many times should diabetics test? Enough tests are needed to give the patient confidence about control or to make appropriate adjustments to treatment. Remember, too many tests, particularly in the very young, may put off insulin diabetics. Initially they should carry out blood tests before main meals and before going to bed; occasionally one or two hours after a main meal to see how high the glucose level rises, and at 3–4 am in patients who are prone to nocturnal hypoglycaemia.

Once control is stable, occasional tests obtained at one of these times or alternatively a full profile of four to six tests obtained two days a week is satisfactory.

For older patients who are not on insulin, blood testing is helpful but not as important. They are less likely to have large, erratic swings in the blood glucose and an occasional fasting test obtained just before breakfast is all that is necessary. A value of less than 7 mmol/l implies good control. In this respect, it is of great value for nurses in the community to carry blood testing equipment.

How can insulin dependent diabetics use blood tests to make adjustments to insulin treatment? Different hospital clinics will have their own routine way of making adjustments in a patient on a mixture of short (clear) and medium (cloudy) acting insulin. Here is an example:

- If the blood test before lunch shows 13 mmol/l, three days in a row, increase the morning short-acting insulin by two units.
- If the test before tea shows 13 mmol/l for more than two days, increase the morning dose of medium-acting insulin by 2–4 units.
- If the bedtime test is high, increase the evening short-acting insulin.
- If the breakfast test is high, increase the

evening medium-acting insulin.

In each case, keep adjusting the dose at two-day intervals until the appropriate blood test is in the range 4–7 mmol/l. If regular hypoglycaemic attacks occur, for instance mid-morning, then the appropriate insulin can be reduced.

At times of illness tests should be performed regularly. If the blood test shows over 13 mmol/l give an extra four units of short-acting (clear) insulin. Repeat the blood test at two-hour intervals and if necessary the dose of four units of short-acting insulin until the blood glucose level falls below 10 mmol/l. When the blood tests improve, the insulin dose can be reduced gradually. In a patient who is on tablets or dietary treatment only for diabetes, it is wise to check the blood test every day. If the results are unsatisfactory, contact the doctor.

It is worth remembering that the test done after a hypoglycaemic attack will often be high – a patient will take extra sugar and his body will react to combat the hypoglycaemia.

Patients can feel hypoglycaemic even when blood glucose is normal – patients who are poorly controlled and used to high blood glucose levels may experience symptoms of a hypoglycaemic attack when the blood glucose falls quickly into the normal range.

Most people can obtain blood from the finger pulp or at the side of the pulp where it is less sensitive. Hands should be washed in warm water for routine cleanliness, but also to remove traces of sugary food from the fingertips. Spirit swabs should be avoided as they will make the skin hard with constant use. Unilet or Monolet lancets will usually produce a drop of blood easily. These lancets can be re-used several times with safety if the protective cap is replaced immediately after use. One of the commonest causes of error in blood testing is obtaining too little blood on the test pad,[5] as smearing and drying rapidly will invalidate the test. For people who find difficulty in pushing in a lancet quickly, one of several spring-loaded finger-pricking devices can be used. Autolet, Autoclix, Autolance and Monotector are all efficient and reasonably priced.

To use testing strips the blood is placed on a pad on the test strip. This is impregnated with enzymes which react with the blood to produce a colour change. For most purposes comparison of this colour change with the scale on the container is accurate enough. However, some have difficulty in comparing shades and prefer to use a special reflectance meter for a reliable estimation. During pregnancy the use of a 'meter' is imperative since very strict standards of glucose control are necessary. Inaccuracy in timing is probably the commonest error and particularly the timing of the first period when blood is placed on the pad. Problems can occur due to degradation of the enzymes in the test pad if it is kept too long after opening or in a particularly warm temperature. This will cause false low readings.

There is probably no best strip. Patients and clinics will have their own preferences depending on cost and reliability.[6,7,8,9] The strips most commonly in use are BM Test-Glycemie 1-44, Visidex II and Glucostix.

A variety of meters is now available. They are usually reliable and easy to operate. There is probably little to choose between them in terms of accuracy[10] and all are compact and robust. Their main disadvantages are that they are expensive (usually £60–£100) and need to be rechecked regularly for accuracy. Many clinics are able to provide meters on loan to try out and they should be available for monitoring during pregnancy. Some newer machines incorporate a memory device and even a mini computer which can be very useful for retaining and analysing results. However, they are expensive and 'gimmicky' for routine use.

The most recently available meters Glucometer II, Reflolux II, Glucocheck TMSc and Hypocount MX are regularly advertised in the British Diabetic Association magazine *Balance*.

Many patients may go through long periods when most tests appear satisfactory and they are tempted to neglect and take results for granted. You should try to encourage a regular routine so that patients can gain confidence and assurance in blood sugar control. They may feel fed up with their testing at times – it is natural that they should. They may be tempted to record good test results to keep the clinic happy. It is vital that clinic staff attempt to understand the difficulties of being diabetic and provide the encouragement and trust to talk openly and honestly. Gaining confidence in using blood tests is a gradual process which can be supported and prompted by everybody who works in the clinic. For patients who become expert in assessment of their diabetic control it can lead to an independent and healthy life with diabetes.

REFERENCES

1 Bodansky, H.J. Tales from the Diabetic Clinic or Why some patients do not bring a record book to the clinic. Practical Diabetes 1986; 3: 247-248

2 Wing, R.R., Lamparski, D.M., Zaslow, S., Betschart, J., Siminerio, L., Becker, D. Frequency and accuracy of self-monitoring of blood glucose in children – relationship to glycaemic control. Diabetes Care 1985; 8: 214-218

3 Scobie, I.N., Bowden, V.K., Lowy, C. Tey, Beng Key, Judd, S.L., Sonksen, P.H.A.

Consumer-orientated survey of long-term home blood glucose monitoring. Practical Diabetes 1985; 2: 33–35.

4 Jones, H., Barman, A.D.E.B., Tattersall, J., Child, D.F. Practical aspects of finger pricking. Practical Diabetes 1985; 2: 26–27.

5 Wing, R.R., Koeske, R., New, A., Lamparski, D., Becker, D. Behavioural skills in self-monitoring of blood glucose: relationship to accuracy. Diabetes Care 1986; 9: 330–33.

6 Erant, D.P., Elkeles, R.S. Visidex II and BM 20-800 Glycemiestrips: a comparison. Practical Diabetes 1985; 2: 40–41.

7 Brown, E., Jones, H., Child, D.F., Pal, G.S., Hudson, P.R. Routine ward BG monitoring by nurses using BM-Test-Glycemie 20-800R strips. Practical Diabetes 1986; 3: 257.

8 Round, J.M., Haseler, M., Sanders, C., Dickson, A., Patel, V., Betteridge, D.J. Asssessment of Visidex II; a new blood glucose monitoring system. Practical Diabetes 1984; 1: 51–154.

9 O'Connell, M., Cregan, D., Keenan, P., Callinan, M., Drury, R., Ivodrury, M. An assessment of field methods for measuring blood and urine glucose levels. Practical Diabetes 1984; 1: 46–48.

10 Petranyi, G., Kyne, D.A., Alberti, K.G.M.M. Evaluation of our modern home blood glucose meters including two memory meters. Diabetic Medicine 1986; 3: 198–190.

Ray Newton FRCPE is consultant physician at Ninewells Hospital in Dundee.

FURTHER READING

PLUMMER, D T (1989) *Biochemistry. The chemistry of life* London, McGraw Hill.

This is a text intended for undergraduate students who need a background in biochemistry. It assumes more knowledge of the underlying sciences than this introductory unit, but can be read by non scientists as it gives a very clear account of the basic principles.

KEELE, C A et al (1982, thirteenth edition) *Samson Wright's Applied Physiology* Oxford, Oxford University Press.

A major medical text. The biochemistry sections are clear and concise, linking the subject to physiology and to observable symptoms. Recommended as a reference book.

'Diabetes Facts Pack', *Community Outlook* May 1992, p. 26.

A useful update, with references.

CARR E K and KENNEY F D (1992) *Positioning the stroke patient: a review of the literature*, International Journal of Nursing Studies 29 (4), 335–369.

A valuable overview of the current thinking and debate about current practice.